Study Guide with Laboratory Activities

Electricity and Electronics

Richard M. Roberts
Assistant Principal for Technology
Tampa Bay Technical Center

Howard H. Gerrish

William E. Dugger, Jr.
Director
Technology for All Americans Project

Publisher
The Goodheart-Willcox Company, Inc.
Tinley Park, Illinois

To my mother Violet, who gave me a strong work ethic.
And to my wife Michele, whose support and
editorial suggestions made this book possible.

Introduction

The ***Electricity and Electronics Study Guide with Laboratory Activities*** is designed to stimulate your interest in electricity and electronics. This guide will help you learn the most important information in the text.

The first *Student Activity Sheet(s)* in each guide chapter review the important concepts, terms, and formulas you learned in the corresponding text chapter. The following *Student Activity Sheets* present laboratory activities that will further your understanding of electronics and should prove very enjoyable.

Each laboratory activity begins with a short introduction that explains the theory behind the activity. This is followed by a list of all the equipment you need to complete the activity. Following the list of equipment are step-by-step instructions that take you through the activity. During these activities, you will be setting up circuits, collecting data, and answering questions. Working with electricity can be dangerous. Always have your circuits inspected by your instructor prior to applying power.

Table of Contents

Chapter 5 Sources of Electricity

Chapter 6 Series Circuits

Chapter 7 Parallel Circuits

Chapter 8 Combination Circuits (Series-Parallel)

Chapter 9 Magnetism

Chapter 10 Generators

Chapter 11 DC Motors

Chapter 12 Transformers

Chapter 13   AC Motors

Chapter 14   Inductance and RL Circuits

Chapter 15   Capacitance and RC Circuits

Chapter 16   Tuned Circuits and RCL Networks

Chapter 17   Introduction to Semiconductors and Power Supplies

Chapter 18   Tubes, Transistors, and Amplifiers

Chapter 19   Integrated Circuits

Chapter 20　Digital Circuits

Chapter 21　Oscillators

Chapter 22　AM and FM Radio Communications

Chapter 23　Television

Chapter 24　Fiber Optics and Lasers

Chapter 25　Computers

Chapter 26　Career Opportunities in Electronics

Materials and Equipment List

Materials and Equipment List

The following list contains all the equipment needed to perform the activities in this guide. In some cases, different components can be substituted for the ones listed. Check the *Materials and Equipment* list for each laboratory activity.

Power Supplies

(1)—0–18 V variable dc supply
(1)—9 Vdc dual voltage power supply (or two 9 V batteries)
(1)—12 Vac power supply
(1)—120 Vac power supply
(2)—6 V batteries
(4)—D cells (good) and holders
(4)—D cells (bad)
(1)—signal generator

Meters and Test Equipment

(2)—multimeters
(1)—digital voltmeter
(1)—galvanometer
(2)—meter leads with alligator clips
(1)—dual-trace oscilloscope with two probes
(1)—logic probe
(1)—electroscope
(1)—breadboard

Resistors and Potentiometers

(1)—10 Ω resistor, 10 W (brown, black, black)
(1)—47 Ω resistor, 1 W (orange, violet, black)
(1)—100 Ω resistor, 1/4 W (brown, black, brown)
(2)—120 Ω resistors, 1/4 W (brown, red, brown)
(1)—200 Ω resistor, 1/4 W (brown, red, red)
(2)—220 Ω resistors, 1/4 W (red, red, brown)
(1)—220 Ω resistor, 1 W (red, red, brown)
(1)—240 Ω resistor, 1/4 W (red, yellow, brown)
(1)—330 Ω resistor, 1/4 W (orange, orange, brown)
(7)—390 Ω resistors, 1/4 W (orange, white, brown)
(4)—510 Ω resistors, 1/4 W (green, brown, brown)
(8)—1 kΩ resistors, 1/4 W (brown, black, red)
(1)—1.2 kΩ resistor, 1/2 W (brown, red, red)
(2)—2.2 kΩ resistors, 1/4 W (red, red, red)

(1)—4.7 kΩ resistor, 1/4 W (yellow, violet, red)
(3)—10 kΩ resistors, 1/4 W (brown, black, orange)
(1)—22 kΩ resistor, 1/4 W (red, red, orange)
(1)—33 kΩ resistor, 1/4 W (orange, orange, orange)
(2)—47 kΩ resistors, 1/4 W (yellow, violet, orange)
(2)—100 kΩ resistors, 1/4 W (brown, black, yellow)
(1)—470 kΩ resistor, 1/4 W (yellow, violet, yellow)
(2)—1 MΩ resistors, 1/4 W (brown, black, green)
(2)—10 MΩ resistors, 1/4 W (brown, black, blue)
(1)—5 kΩ potentiometer
(1)—10 kΩ potentiometer
(1)—50 kΩ potentiometer
(1)—100 kΩ potentiometer

Capacitors

(1)—1 pF capacitor, 50 V
(1)—4 pF capacitor, 50 V
(1)—5 pF capacitor, 50 V
(2)—47 pF capacitors, 50 V
(1)—470 pF capacitor, 50 V
(2)—0.01 μF capacitors, 50 V
(2)—0.022 μF capacitors, 50 V
(1)—0.1 μF (100 pF) capacitor, 50 V
(1)—1 μF capacitor, 50 V
(1)—6 μF capacitor, 50 V
(1)—2.2 μF capacitor, 50 V
(1)—3.3 μF capacitor, 50 V
(1)—4.7 μF capacitor, 50 V
(2)—10 μF capacitors, 50 V
(1)—33 μF capacitor, 50 V
(1)—47 μF capacitor, 50 V
(2)—100 μF capacitors, 50 V
(1)—200 μF capacitor, 50 V
(1)—1000 μF capacitor, 35 V
(1)—variable capacitor, 360 pF

Inductors and Transformers

(1)—RF choke, 100 μH
(1)—center tap transformer, 120 V primary, secondary 6 V–6 V, 1.2 A
(1)—center tap transformer, 120 V primary, secondary 6.3 V–6.3 V, 3 A
(1)—audio output transformer

Diodes, Transistors, and ICs

(2)—diodes #1N4001
(4)—diodes #1N4005
(1)—diode #1N5402
(1)—diode #1N34A (germanium)
(1)—diode #1N4735 (zener, 6.2 V)
(2)—LEDs (red)
(1)—LED (yellow)
(1)—LED (green)
(8)—LEDs (any color)
(1)—LED seven segment display, common cathode

(1)—SCR

(2)—NPN transistors, MPS2222A (or equivalent)

(1)—NPN transistor, MPS3904

(1)—NPN switching transistor

  silicon, $h_{FE} = 200$

  $I_C = 800$ mA, $V_{CE} = 30$ V

  power dissipation = 1.8 W

(10)—assorted transistors

(1)—4001 quad 2-input NOR gate

(1)—4011 quad 2-input NAND gate

(2)—7404 hex inverters

(1)—7408 quad 2-input AND gate

(1)—7432 quad 2-input OR gate

(1)—7447 decoder/driver

(2)—7476 J-K flip-flops

(1)—74HCT86 quad 2-input exclusive OR gate

(1)—74HCT393 ripple counter

(1)—555 timer

(1)—741 op amp

(1)—LM386 audio amplifier

Additional Electric and Electronic Equipment

(2)—dc motors, 9–18 V

(1)—single-phase motor, 1/4 hp (Baldor or equivalent)

(1)—lamp, 6 Vdc and holder

(4)—lamps, 12 Vdc and holders

(1)—lamp, 14 Vdc and holder

(1)—fuse, 1/4 A and holder

(4)—SPST switches

(5)—SPDT switches

(2)—DPDT switches

(2)—push-button switches, NO

(2)—push-button switches, NC

(1)—momentary contact push-button switch

(1)—SPDT relay, 12 Vdc

(2)—DPDT relays, 12 Vdc

(1)—test pigtail, 120 V

(1)—coaxial converter, 75 Ω

(1)—speaker, 8-ohm

(1)—piezoelectric speaker element

(1)—condenser microphone

(1)—earphone

(1)—GP1U52X infrared receiver/demodulator

(1)—pair infrared emitter and detector

(1)—photocell

(2)—silicon solar cells

(1)—portable AM/FM radio

(1)—television

(1)—remote control transmitter (typical TV remote)

(1)—IBM PC or compatible

(1)—MS DOS Software System (standard with most computers) with QBasic.Exe

Wire

—18 inches of #14 AWG bare solid conductor
—supply of #22 connection wire
—120 feet of #30 AWG magnet wire
—selection of colored jumper wires
—two-wire ribbon flat lead

Miscellaneous Supplies

(1)—2″ × 2″ square of aluminum foil
(1)—2″ piece of drinking straw
(1)—20d nail
(1)—3″ × 5″ card
(6)—tablespoons of salt
(1)—antenna support
(1)—beaker
(1)—blue pen or pencil
(1)—compass
(1)—drinking straw
(1)—glass rod and silk
(1)—good grounding source (2 feet of rod or pipe)
(1)—length of heavy conductor or 1/2″ or smaller tubing
(1)—marker
(1)—measuring spoon
(1)—pencil or pen
(1)—plastic mallet
(1)—plywood base (8″ × 4″)
(1)—plywood base (12″ × 24″)
(1)—red pen or pencil
(1)—set of insulated screwdrivers
(1)—set of socket drives
(1)—stopwatch (or wristwatch with a second hand)
(1)—vulcanite rod and fur
(1)—watch or clock
(4)—conduit straps
    —12–15 oz. tap water
    —4 inches of 3/4 inch PVC pipe
    —supply of scratch paper
    —supply of tape
    —various types of metal plates:

Copper	Aluminum
Iron	Silver
Lead	Zinc

Science of Electricity and Electronics

Student Activity Sheet 1-1

Review

Name _____ **Score** _____

Date _____ **Class/Period/Instructor** _____

Complete the following sentences by filling in the missing word or words.

1. A compound is composed of two or more _____.

   1. _____

2. The smallest part of an element is called a(n) _____.

   2. _____

3. _____ are in orbit around the nucleus of an atom.

   3. _____

4. An electron is considered to have a(n) _____ charge; a proton has a(n) _____ charge; a neutron exhibits _____ charge.

   4. _____
   _____
   _____

5. The nucleus of an atom is composed of _____ and _____.

   5. _____
   _____

6. When an atom loses an electron it is said to be _____.

   6. _____

7. An atom with an extra electron becomes a(n) _____ ion.

   7. _____

8. An atom with one less electron than proton becomes a(n) _____ ion.

   8. _____

9. _____ charges repel each other, while _____ charges attract each other.

   9. _____
   _____

10. There are _____ electrons in one coulomb.

    10. _____

11. The force surrounding a charged body is called a(n) _____ field or a(n) _____ field.

    11. _____
    _____

12. Charging an object by placing a charged body near it is an example of _____.

    12. _____

13. Pollution and small particles can be removed from the air by using a(n) _____ _____.

    13. _____

(Continued)

14. A basic electrical circuit consists of three parts. Name them.

14. _____

_____

_____

15. An excess of electrons produces a(n) _____ charge, while a depletion of electrons produces a(n) _____ charge.

15. _____

_____

16. The plus sign (+) represents a(n) _____ connection point. A negative sign (–) represents a(n) _____ connection point.

16. _____

_____

17. When you are uncertain of how to connect an electrical device, you should _____ _____ _____.

17. _____

18. The force of electron flow is measured in _____. The number of electrons flowing is measured in _____. The opposition to electron flow is measured in _____.

18. _____

_____

_____

19. Voltage is expressed by the letter _____. The volt is abbreviated with a(n) _____.

19. _____

_____

20. Current is expressed by the letter _____. The ampere is abbreviated with a(n) _____.

20. _____

_____

21. Resistance is expressed by the Greek symbol _____.

21. _____

22. _____ _____ flows in one direction, while _____ _____ flows back and forth through a conductor.

22. _____

_____

23. _____ _____ _____ theory states that electricity flows from positive to negative. _____ _____ theory states that electricity, or the electron, flows-from negative to positive.

23. _____

_____

24. A circuit that provides only one path for electrons to flow through is called a(n) _____ circuit. A circuit that provides more than one path for electron flow is called a(n) _____ circuit.

24. _____

_____

25. The relationship of voltage, current, and resistance can be calculated by applying _____ _____.

25. _____

26. Voltage is equal to _____ multiplied by _____.

26. _____

_____

27. Current is equal to _____ divided by _____.

27. _____

_____

(Continued)

Fundamentals of Electricity and Electronics

28. Resistance is equal to _____ divided by _____.

28. _____

_____

29. A 24-volt battery will produce _____ amperes when connected to a 6-ohm resistance.

29. _____

30. There are _____ millivolts in 4 volts.

30. _____

Static Electricity Experiment

Name _____ **Score** _____

Date _____ **Class/Period/Instructor** _____

Introduction

Static electricity is produced by friction. In this lab, you will explore common properties of static electricity. You will produce a static charge and record its effect on the electroscope. You will be able to describe the effects of induction.

Materials and Equipment

(1)—electroscope

(1)—glass rod and silk

(1)—vulcanite rod and fur

Electroscope

Procedure

Step 1. Gather all materials required for this activity.

Step 2. Make sure the electroscope is fully discharged by touching the ball at the top.

Step 3. Briskly rub the vulcanite rod with the fur. Place the tip of the rod near the top of the electroscope, being careful not to touch the top with the glass rod. Then slowly remove it.

Question 1. Describe what you observed happened to the leaves of the electroscope.

_____

_____

_____

Question 2. Would you say that the action proved that like charges attract or repel? Why?

_____

_____

_____

Step 4. Rub the vulcanite rod again. This time, touch the tip of rod to the top of the electroscope. Remove the rod.

Question 3. Describe what happened when you touched the top and then withdrew the rod this time.

_____

_____

_____

(Continued)

Step 5. Discharge the electroscope and repeat the procedures in Step 3. This time use the glass rod and the silk.

Question 4. Describe what happened when you placed the glass rod near the tip of the electroscope this time.

_____

_____

_____

Step 6. Repeat the procedures in Step 4 using the glass rod and silk.

Question 5. Describe what happened when you touched the glass rod to the tip of the electroscope.

_____

_____

_____

Step 7. Charge either rod and hold it in your *right* hand. Bring your *left* hand near the top of the electroscope. Record your findings and explain.

Question 6. Describe what happened when your hand came near the top of the electroscope.

_____

_____

_____

Question 7. When you charge either rod, is it possible to determine the polarity? If so, state how.

_____

_____

_____

Step 8. Clear your work area. Properly store equipment and supplies.

Fundamentals of Electricity and Electronics

Ohm's Law Practice

Name_____ **Score**_____

Date_____ **Class/Period/Instructor**_____

The only way to master Ohm's law is with practice. The following problems are simple to solve. The challenge of this exercise is to become familiar with how voltage, current, or resistance can be solved for when two factors are known. All three forms of the Ohm's law formula are provided below.

$E = I \times R$ Used to find *voltage (E)* when current (I) and resistance (R) are known.
$R = E/I$ Used to find *resistance (R)* when voltage (E) and current (I) are known.
$I = E/R$ Used to find *current (I)* when voltage (E) and resistance (R) are known.

1. _____ volts   2 amperes   5 ohms

2. _____ ohms   6 volts   3 amps

3. _____ amps   12 volts   2 ohms

4. _____ ohms   12 volts   5 amps

5. _____ volts   5 amps   10 ohms

6. _____ amps   10 ohms   24 volts

7. _____ ohms   5 amps   6 volts

8. _____ volts   0.25 amps   12 ohms

9. _____ ohms   1.5 amps   6 volts

10. _____ amps   100 ohms   12 volts

11. _____ volts   25 ohms   3 amps

12. _____ volts   0.06 amps   1200 ohms

13. _____ volts   0.01 amps   122 ohms

14. _____ ohms   0.005 amps   6 volts

15. _____ volts   1800 ohms   0.05 amps

16. _____ volts   1400 ohms   0.007 amps

17. _____ amps   18 volts   3600 ohms

18. _____ amps   24 volts   600 ohms

19. _____ amps   12 volts   1200 ohms

20. _____ volts   16 ohms   2 amps

Common Electrical Terms and Symbols

Name_____ **Score**_____

Date_____ **Class/Period/Instructor**_____

In this exercise, you will become familiar with the common terms and symbols associated with basic electrical theory. This exercise also provides you with more practice using Ohm's law. The following electrical terms and symbols are used to describe electrical circuit action.

Voltage	E, V, volts, electrical pressure, electromotive force (emf), potential
Current	I, A, amperes, amps, electron flow
Resistance	R, Ω, ohms

Relate the terms and symbols to Ohm's law, and solve for the unknown values. Formulas for Ohm's law can be found on the previous Student Activity Sheet or in the textbook.

1. _____ volts   3 Ω 5 A

2. _____ Ω 12 E 6 A

3. _____ A   6 emf   24 Ω

4. _____ potential   20 R 2 A

5. _____ resistance   0.25 A   12 V

6. _____ amperes   100 Ω 24 E

7. _____ E 1000 Ω 0.01 A

8. _____ ohms   18 V   0.025 amperes

9. _____ V   2200 Ω 0.05 I

10. _____ I 6 volts   100 ohms

Applying Ohm's Law, Electrical Terms, and Symbols Using Word Problems

Name_____ **Score**_____

Date_____ **Class/Period/Instructor**_____

In electrical work and experimentation, the technician must communicate effectively with other technicians. By successfully comprehending the following problems and solving for the unknown values, you will be able to improve your communication skills and raise your technical reading level. When needed, you can round off your answers to an accuracy of three significant digits.

1. A meter should indicate _____ volts when connected across a 100 Ω resistor that has 6 amps of current passing through it.

    1._____

2. A toaster with a 20-ohm heater element would produce a current of _____ A when connected to a 120-volt circuit.

    2._____

3. A 12-volt indicator lamp would have a resistance value of _____ ohms to produce a current of 0.5 amps.

    3._____

4. A voltage of _____ V is needed to produce a 1.5-ampere current reading when connected to a 20-ohm resistance.

    4._____

5. _____ amperes will be present when a 33-ohm resistor is connected to a 6-volt source.

    5._____

6. _____ volts are required to produce 0.005 amps of current through a 100-ohm resistor.

    6._____

7. A(n) _____ Ω resistor will produce a current equal to 0.025 amps when connected to a 12-volt source.

    7._____

8. A 6-volt source connected to a 1200-ohm resistor will cause a(n) _____ amp current.

    8._____

9. A potential equal to _____ V is needed to provide 0.010 amps of current through a 100-ohm resistor.

    9._____

10. A(n) _____ amp current will be produced when a 220-ohm resistor is connected to a 6-volt source.

    10._____

Student Activity Sheet 1-6
Electrical Prefixes

Name _____ **Score** _____

Date _____ **Class/Period/Instructor** _____

    In this activity, you will become familiar with the prefixes commonly used in electrical/electronics trades. The prefixes are important to know and understand since values are often read or spoken using them. For example, a technician may write "40 kV" rather than 40,000 volts.

Prefix	Symbol	Decimal Equivalent	Power of Ten
tera	T	1,000,000,000,000.	10^{12}
giga	G	1,000,000,000.	10^{9}
mega	M	1,000,000.	10^{6}
kilo	k	1,000.	10^{3}
basic unit		1.	
milli	m	.001	10^{-3}
micro	μ	.000 001	10^{-6}
nano	n	.000 000 001	10^{-9}
pico	p	.000 000 000 001	10^{-12}

Express the values on the left as an equal quantity on the right using the electrical prefix provided.

1.   10,000 volts   =   _____ kV

2.   25,000,000 volts =   _____ MV

3.   500 volts   =   _____ kV

4.   0.050 amps   =   _____ mA

5.   1.2 A   =   _____ mA

6.   0.0005 A   =   _____ mA

7.   2000 ohms   =   _____ kΩ

8.   12,500 ohms   =   _____ kΩ

9.   122,000 ohms   =   _____ MΩ

10. 4.5 kΩ   =   _____ MΩ

11. 0.025 mA   =   _____ μA

12. 0.000005   =   _____ μA

13. 0.5 kV   =   _____ volts

14. 25 μV   =   _____ volts

15. 500 μA   =   _____ mA

16. 23 GV   =   _____ kV

17. 2.2 kΩ   =   _____ MΩ

18. 2,000,125 Ω   =   _____ kΩ

19. 3.15 kΩ   =   _____ Ω

20. 50 mA   =   _____ A

Ohm's Law Practice Using Common Electronic Prefix Values

Name_____ Score_____

Date_____ Class/Period/Instructor_____

The problems below are a combination of the Ohm's law and electrical prefix activity sheets (Student Activity Sheets 1-3 and 1-6). Express all answers using the most appropriate prefix.

1. A current of _____ amps will be present when a 24 kΩ resistance is connected to a 12-volt supply.

1. _____

2. A source of _____ volts is required to create a current of 0.125 amps through a 50 kΩ resistance.

2. _____

3. A(n) _____ ohm resistance is needed to create a 50-milliamp current when connected to a 6-volt source.

3. _____

4. When 50 microamps are flowing through a 60 kΩ resistor, a(n) _____ volt source is supplying the potential for the circuit.

4. _____

5. A voltage value of _____ volts is required to exert sufficient force to produce a current of 5 μA through a 20 MΩ resistance.

5. _____

6. A 4 MΩ resistance connected to a 120-volt source will generate a current value equal to _____.

6. _____

7. A source of 1200 mV is applied to a resistance of 3.6 kΩ and produces a current equal to _____.

7. _____

8. When a 24-volt source is connected to an 8 kΩ resistance, a current of _____ will be produced.

8. _____

9. A resistance value of 1.2 kΩ also has a current value equal to 25 mA. There must be a potential equal to _____.

9. _____

10. A resistor value of 3.6 kΩ has a current value equal to 25 μA. This means that there must be a supply of _____ connected to the resistance.

10. _____

11. A 15 kΩ resistor connected to a 12-volt battery will produce _____A.

11. _____

12. _____ Ω are required to produce 15 mA when connected to 1.5 V.

12. _____

13. A 24-volt battery will produce _____ when connected to a 450 kΩ resistance.

13. _____

14. A _____ volt source is required to produce 12 μA through a 240 kΩ resistance.

14. _____

15. An EMF equal to _____ is needed to produce a current of 0.000015 amps when connected to a 1,200,000 ohm resistance.

15. _____

16. A voltage equal to _____ is required to produce 360 kA when connected to a load equal to 2 ohms.

16. _____

17. A generator produces 14.2 kV while supplying a(n) _____ Ω load at 7 amps.

17. _____

(Continued)

18. _____ will be produced when an 800 Ω resistor is connected to a 6-volt source.

18. _____

19. _____ will be required to provide 0.015 amperes when connected to a 10-volt EMF.

19. _____

20. _____ will occur when a 2400-ohm resistor is connected to a 6-volt source.

20. _____

Basic Instruments and Measurements

Student Activity Sheet 2-1
Review

Name_____ **Score**_____

Date_____ **Class/Period/Instructor**_____

Complete each statement below by filling in the missing word or words.

1. A red plus (+) sign on a meter represents _____ polarity.

    1. _____

2. A black negative (–) sign on a meter represents _____ polarity.

    2. _____

3. A(n) _____ is used to measure the volume of electrons flowing through a circuit.

    3. _____

4. An ammeter is connected in _____ with the load it is reading.

    4. _____

5. To extend the range of an ammeter, a(n) _____ can be added. These are connected in _____ to the meter movement coil.

    5. _____

    _____

6. An ammeter can be easily damaged if connected directly across the source voltage because its meter coil winding has a very _____ resistance.

    6. _____

7. A(n) _____ is used to measure electrical pressure in a circuit.

    7. _____

8. A voltmeter is connected in _____ with the circuit.

    8. _____

9. The sensitivity of a voltmeter is expressed in units of _____.

    9. _____

10. To extend the range of a voltmeter, _____ _____ can be added to the meter movement circuit. They are connected in _____ with the meter coil.

    10. _____

    _____

11. A(n) _____ is used to measure the opposition to current in a circuit.

    11. _____

12. A(n) _____ is damaged if the circuit remains energized while taking a reading.

    12. _____

13. An ohmmeter utilizes a(n) _____ scale, which means that the markings on the scale are not equally spaced.

    13. _____

(Continued)

14. A VOM can be used to read _____, _____, and _____.

14. _____

_____

_____

15. The _____ does not use an analog meter movement. It uses a(n) _____ _____ _____ to display the electrical readings.

15. _____

_____

16. An electrical drawing that indicates the location of each component in a circuit and the value of each is known as a(n) _____ diagram.

16. _____

17. An electrical drawing that indicates how components are grouped together to form stages is called a(n) _____ diagram.

17. _____

18. A meter movement coil has a resistance of 50 ohms and a full scale deflection is achieved by 0.001 ampere. The total applied voltage to achieve full scale deflection is _____.

18. _____

19. The meter in question 18 could double its range as a voltmeter by adding a(n) _____ resistor in series with the coil.

19. _____

20. The meter in question 18 could increase its range as a voltmeter tenfold (by 10 times) by adding a(n) _____ resistor in series with the coil.

20. _____

Fundamentals of Electricity and Electronics

Reading Linear and Nonlinear Scales

Name_____ Score_____

Date_____ Class/Period/Instructor_____

A variety of linear and nonlinear scales will be encountered when reading meters. This activity sheet will provide you with practice in reading and interpreting linear meter scales.

1. There are 10 meter readings indicated on the scale below (marked A–J). Record the meter readings by writing the values on the lines provided. Your accuracy should be to at least two decimal places. Take special note of the scale selector switch setting on the right side of the scale before you begin.

    A._____

    B._____

    C._____

    D._____

    E._____

    F._____

    G._____

    H._____

    I._____

    J._____

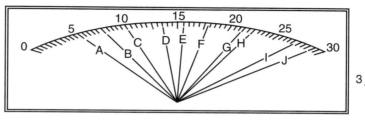

2. There are 10 meter readings indicated below (marked A–J). Record the meter readings by writing the values on the lines provided. Your readings should be accurate to two decimal places. Take special note of the scale selector switch to the right of the scale.

    A._____

    B._____

    C._____

    D._____

    E._____

    F._____

    G._____

    H._____

    I._____

    J._____

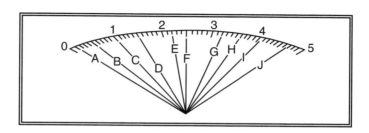

(Continued)

Chapter 2   Basic Instruments and Measurements

3. Below is a nonlinear scale such as ones used for reading resistance on an analog meter. There are 10 meter readings indicated below (marked A–J). Record the meter readings by writing the values on the lines provided. Take special note of the scale selector switch to the right of the scale.

A._____

B._____

C._____

D._____

E._____

F._____

G._____

H._____

I._____

J._____

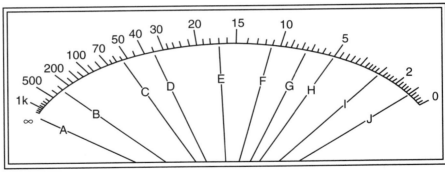

4. Below is a nonlinear scale such as ones used for reading resistance on an analog meter. There are 10 meter readings indicated below (marked A–J). Record the meter readings by writing the values on the lines provided. Take special note of the scale selector switch to the right of the scale.

A._____

B._____

C._____

D._____

E._____

F._____

G._____

H._____

I._____

J._____

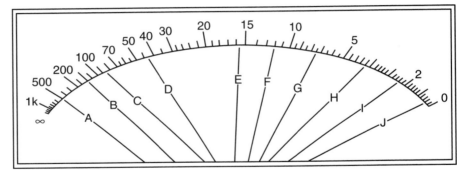

Meter Connection Practice

Name _____ **Score** _____

Date _____ **Class/Period/Instructor** _____

Introduction

It is important that you have an understanding of schematic diagrams. It is also important that you know how to correctly connect meters to a circuit. This practice will assist you in making proper meter connections and introduce you to properly interpreting schematic diagrams.

Procedure

Select the schematic that best meets the condition stated and place the corresponding letter in the blank provided in the statement.

_____ 1. Represents an ammeter properly connected to read the current through the resistor.

_____ 2. Represents a voltmeter properly connected to read the voltage at the resistor.

_____ 3. Represents an ohmmeter connected properly to read the resistance value of the resistor.

_____ 4. Represents a connection that will result in damage to the ammeter. Why will this connection result in damage to an ammeter?

_____

_____

_____

_____

Student Activity Sheet 2-4

Power Supply Familiarization

Name_____ Score_____

Date_____ Class/Period/Instructor _____

Introduction

In many lab experiments, you will need to work with a power supply in order to energize your circuits. This activity will familiarize you with the power supply and allow you to experiment with it in preparation for future Student Activity Sheets.

Materials and Equipment

(1)—power supply

Procedure

You will answer questions about the power supply to verify that you know the purpose of each part. Correct responses will ensure the necessary knowledge of the power supply parts. Incorrect answers will be corrected by your instructor to ensure a thorough understanding of your power supply.

Meter Scales

Question 1. Is your power supply equipped with meter scales? _____

Question 2. What does the meter scale(s) measure? _____

_____

Overload Protection

Question 3. Does your power supply provide overload protection?_____

Question 4. What type of overload protection (fuse, breaker, reset, internal, etc.) is used? _____

_____

Voltage

Question 5. What type of voltages (dc, ac, variable dc, variable ac) are available and what are the maximum

values? _____

_____

_____

_____

Question 6. What color are the dc voltage terminals? _____

Question 7. What color are the ac voltage terminals? _____

Question 8. Are there lights on the face of the meter indicating that power is present and/or that overload

conditions exist? _____

Question 9. How is the power supply turned on? _____

Question 10. On a separate sheet of paper, sketch the face of your power supply and properly label the main parts. Be sure to include meter scales, voltage adjustment, and output terminals.

Student Activity Sheet 2-5

Verifying Ohm's Law Current and Voltage Relationship

Name _____ **Score** _____

Date _____ **Class/Period/Instructor** _____

Introduction

This activity will teach you the proper methods of connecting and using voltmeters and ammeters. Construct a graph representing the relationship of current and voltage when applied to a fixed resistance. You will calculate the expected value using Ohm's law, and then take a reading using the ammeter. The circuit characteristics of many electronic components are explained using graphs. It is important that you understand how graphs are constructed using electrical terms.

Materials and Equipment

(1)—1–12 volt variable power supply

(1)—breadboard

(1)—220 Ω resistor, 1 W

(1)—#22 connection wire

(1)—ammeter

(1)—voltmeter

(1)—ohmmeter

(1)—red pen or pencil

(1)—blue pen or pencil

Procedure

Step 1. Gather all materials required for this activity.

Step 2. Check that the power supply is off. Be sure that it is turned off before you begin to construct the circuit.

Step 3. Measure the resistance of the 220-ohm resistor with the ohmmeter. Record the value below. Refer to it when calculating the values for the chart marked "Actual Current Value."

Actual Resistance Value = _____ ohms

Step 4. Connect the circuit as indicated in the following schematic.

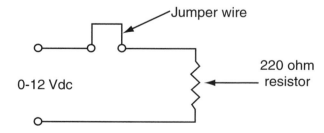

The jumper is where you will connect the ammeter in the circuit. Remember that the ammeter must be in series with the resistor to obtain a current reading. If the ammeter is connected in parallel, it may be damaged.

Chapter 2   Basic Instruments and Measurements

(Continued)

Step 5. Connect the voltmeter in parallel to the resistor as indicated in the following schematic diagram. Do *not* turn on the power supply until your instructor has checked your circuit.

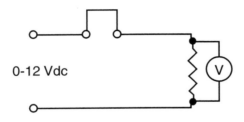

0-12 Vdc

Step 6. Turn on the power supply, and adjust the output to 2 volts dc.

Step 7. Turn off the power supply. Disconnect the voltmeter. Remove the jumper wire and connect the ammeter as indicated in the schematic that follows. Have your instructor check the connections. Do *not* turn the power supply on until your instructor has approved your ammeter connection. The ammeter can be easily damaged if improperly connected.

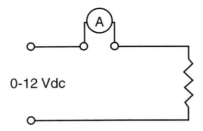

0-12 Vdc

Step 8. Turn on the power supply, and record the current value on the chart marked "Actual Current Value."

Step 9. Repeat the previous steps, increasing the power supply voltage by two volts each time. Record the current values on the chart until all values are complete from 2 volts to 12 volts. Have your instructor check your circuit each time to be sure that the meters are properly connected.

Applied Voltage Value	Actual Resistance Value	Calculated Current $I = E/R$	Actual Current Value
2			
4			
6			
8			
10			
12			

(Continued)

Fundamentals of Electricity and Electronics

Step 10. After all values have been entered in the chart, proceed to plot the values for actual current in red on the graph provided below. Plot calculated current value in blue.

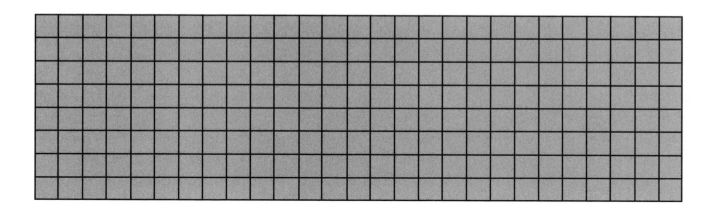

Step 11. In your own words, write a conclusion for the experiment. Report if the relationship of current and voltage is directly or indirectly related. If the relationship is directly related, the current will increase with an increase of voltage. If the relationship is inversely related, the current will decrease with an increase of voltage.

_____

_____

_____

_____

_____

_____

_____

_____

Step 12. Clear your work area. Properly store equipment and supplies.

Meter Loading

Name _____ **Score** _____

Date _____ **Class/Period/Instructor** _____

Introduction

This lab activity will demonstrate the effects of meter loading. The electrical meters used to take readings such as voltage actually become part of the circuit. By becoming part of the circuit and providing another path for current, the characteristics of the circuit being evaluated may change.

Materials and Equipment

(1) — voltmeter

(2) — 10 kΩ, 1/4 W resistor (brown, black, orange)

(2) — 10 MΩ, 1/4 W resistor (brown, black, blue)

(1) — SPST switch

(1) — breadboard

(1) — 12 Vdc power supply

Procedure

Step 1. Gather all materials required for this activity.

Step 2. Assemble a circuit consisting of two 10 kΩ resistors connected in series. (Use the schematic below to assist you.) Do *not* energize the circuit until your instructor checks your circuit.

Step 3. Close the SPST switch and energize the circuit. Connect the digital voltmeter across the power supply and record the exact voltage.

Power supply voltage = _____ volts

Step 4. While the circuit is energized, read the voltage drop across R_1 and then R_2. Record the voltage readings.

R_1 voltage drop = _____ volts

R_2 voltage drop = _____ volts

Question 1. When added, the two voltage drops at R_1 and R_2 should equal the source voltage. Do they match, or are they close? _____

Step 5. Turn off the power supply and reconstruct the circuit using the 10 MΩ resistors in place of the 10 kΩ resistors. Energize the circuit, and once again take voltage drop readings across the two resistors. Record the readings, and compare them to the 10 kΩ resistor readings.

R_1 voltage drop = _____ volts

R_2 voltage drop = _____ volts

(Continued)

Question 2. Did the voltage drops add up to equal the source voltage this time? If not, how far off were the readings? _____

_____

_____

Step 6. Turn off the power. Clear your work area. Properly store equipment and supplies.

Fundamentals of Electricity and Electronics

Introduction to Basic Electrical Circuit Materials

Student Activity Sheet 3-1

Review

Name_____ **Score**_____

Date_____ **Class/Period/Instructor**_____

Complete each statement below by filling in the missing word or words.

1. Conductors with a large diameter can carry _____ (more/less) current than small-diameter conductors of the same material.

   1. _____

2. The circular mil system is based on the _____ of the conductor.

   2. _____

3. A mil is equal to _____ inch.

   3. _____

4. There are _____ mils in 1/8 inch.

   4. _____

5. The circular mil area of a conductor that has a diameter of 1/8 inch is _____.

   5. _____

6. The formula for circular mil area is _____.

   6. _____

7. The circular mil area of a #28 conductor is _____.

   7. _____

8. List the four factors that affect the resistance of a conductor.

   _____

   _____

   _____

   _____

9. As the cross-sectional area of a conductor increases, its resistance _____.

   9. _____

10. As a conductor length increases, so does its _____.

    10. _____

11. As the temperature of a copper conductor increases, its resistance _____.

    11. _____

12. Electrical pressure loss across a length of conductor is called _____.

    12. _____

13. The formula for computing the voltage drop of a conductor is _____.

    13. _____

14. A 1000-foot piece of #22 copper conductor has a resistance of _____ ohms.

    14. _____

15. The abbreviation SPST means _____.

    15. _____

16. Two forms of circuit protection are the _____ and the _____.

    16. _____

    _____

(Continued)

17. A(n) _____ is a reusable form of circuit protection.

18. A lamp that has no filament is referred to as a(n)_____ _____ _____.

19. When comparing the quantity of light a lamp produces, you would compare _____ per watt.

20. Resistors are rated in _____ and _____.

17. _____

18. _____

19. _____

20. _____

_____

Converting Resistor Color Code

Name_____ **Score**_____

Date_____ **Class/Period/Instructor**_____

Technicians must be able to convert resistance values from color bands to equivalent resistance values. Indicate the resistance value, maximum resistance, and minimum resistance to be expected for each of the following color combinations.

Color Bands	Resistance Value (Ω)	Maximum Value (Ω)	Minimum Value (Ω)
Example: Red, Blue, Red, Gold	*2600Ω*	*2860Ω*	*2340Ω*
1. Red, Yellow, Red			
2. Brown, Blue, Orange			
3. Green, Red, Brown			
4. Gray, Green, Black, Gold			
5. Blue, Brown, Red, Silver			
6. Green, Red, Silver			
7. Red, Blue, Gold			
8. Brown, Brown, Brown			
9. Brown, Red, Silver			
10. Violet, Gray, Red, Gold			
11. Yellow, Orange, Black			
12. Blue, Gray, Red			
13. Green, Violet, Orange			
14. Blue, Red, Silver			
15. Yellow, Brown, Red			
16. Brown, Black, Black			
17. Brown, Red, Gold			
18. Gray, Blue, Red			
19. Red, Brown, Gray			
20. Violet, Red, Brown			

Matching Resistor Color Code

Name_____ **Score**_____

Date_____ **Class/Period/Instructor** _____

Match the numerical values with their color codes. In the blank spaces provided, write the letter from the right column with the color sequence that represents the resistor value in the left column.

1._____	12,000 Ω	A.	Red, Red, Gold
2._____	22 kΩ	B.	Yellow, Violet, Yellow
3._____	56 kΩ	C.	Brown, Red, Orange
4._____	470 kΩ	D.	Gray, Red, Silver
5._____	1.5 kΩ	E.	Blue, Gray, Red
6._____	180 Ω	F.	Brown, Black, Brown
7._____	2.2 Ω	G.	Brown, Black, Black
8._____	3.3 kΩ	H.	Brown, Black, Red
9._____	54 kΩ	I.	Yellow, Violet, Black
10._____	1.5 MΩ	J.	Brown, Green, Red
11._____	390 Ω	K.	Red, Red, Orange
12._____	1 kΩ	L.	Gray, Red, Brown
13._____	4.7 Ω	M.	Brown, Green, Green
14._____	820 Ω	N.	Brown, Gray, Brown
15._____	1.2 kΩ	O.	Green, Blue, Orange
16._____	100 Ω	P.	Orange, Orange, Red
17._____	0.82 Ω	Q.	Brown, Red, Red
18._____	10 Ω	R.	Yellow, Violet, Gold
19._____	47 Ω	S.	Orange, White, Brown
20._____	6.8 kΩ	T.	Green, Yellow, Orange

Student Activity Sheet 3-4

Switches and Lamps

Name _____ **Score** _____

Date _____ **Class/Period/Instructor** _____

Introduction

Switches and lamps are common devices found in simple circuits. You will use switches and lamps in many electronic/electrical activities. This activity will familiarize you with common switch applications as well as lamps connected in series and parallel.

Materials and Equipment

(1)—12 Vdc variable power supply

(1)—breadboard

(2)—SPST switches

(2)—DPDT switches

(1)—push-button switch, NO

(1)—push-button switch, NC

(3)—No. 47 lamps, 12 V

NO push-button

NC push-button

SPST switch

SPDT switch

DPDT switch

Lamp

Procedure

Step 1. Gather all materials required for this activity. Each circuit will use a 12 Vdc power supply.

Step 2. Connect the circuit shown in the following schematic on your proto board. Have your instructor approve your connections before energizing the circuit. Note the brightness of the lamp.

Step 3. Connect the following parallel circuit. Have your instructor check the circuit before energizing it.

Question 1. After you energized the circuit, did both lamps burn at the same brightness? Did both lamps burn as brightly as the single lamp in Step 2? _____

_____

_____

(Continued)

Chapter 3   Introduction to Basic Electrical Circuit Materials

Step 4. Connect the series circuit according to the following schematic. Have your instructor check it before energizing the circuit.

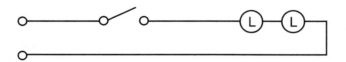

Question 2. Did the lamps connected in series burn at the same brightness as the lamps connected in parallel?

_____

Question 3. Did the two lamps burn in equal brightness to each other in the series circuit?

_____

Step 5. Connect the following circuit and have your instructor check it before energizing. Notice that the circuit consists of a single-pole, double-throw switch this time. *(Use one half of a double-pole, double-throw switch to create the single-pole, double-throw switch.)*

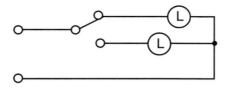

Question 4. Why does one lamp stay on at all times? _____

_____

_____

Step 6. Connect the following circuit. Have your instructor check the circuit before energizing it. This circuit can be easily wired incorrectly so be careful when wiring it.

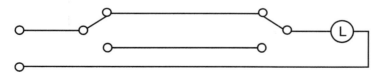

Question 5. Why does either switch have control of the lamp? _____

_____

_____

Step 7. This next circuit allows you to control a lamp from three separate locations. To make this circuit work properly you will need to wire a DPDT switch in the center of the two SPST switches. Take special note of how the conductors cross at the DPDT switch. This is the secret to making the circuit work correctly.

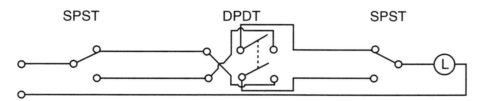

Step 8. Draw a circuit in the space below using four switches to control a lamp from four different locations. After your drawing has been approved by your instructor, proceed to wire it.

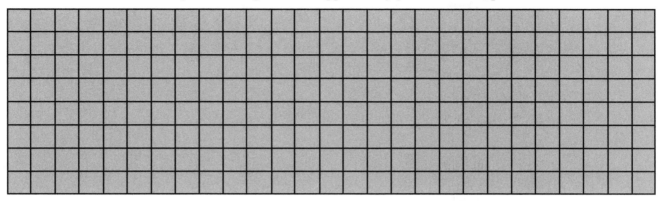

Question 6. What is the maximum number of switch locations you could use to control a lamp using SPST and DPDT switches? _____

_____

Step 9. Connect the following circuit. The NO push-button switch will control only one lamp. The NC push-button switch will control two lamps. Have your instructor check the circuit before you energize it.

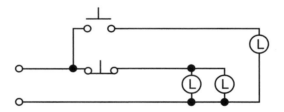

Question 7. How many lamps were lit when the circuit was energized? Why? _____

_____

_____

Question 8. Did all the lamps burn at the same brightness? If so, why? _____

_____

_____

Step 10. Turn off the power supply. Clean up your area, and properly store your equipment.

Converting Circuit Descriptions to Schematics

Name_____ Score_____

Date_____ Class/Period/Instructor_____

Introduction
This activity will provide you with practice in converting descriptions of circuits to schematic diagrams. This activity is important when learning to design your own circuits.

Materials and Equipment
pencil or ink pen

scratch paper

Problems
Problem 1. Draw a circuit consisting of three lamps that can be turned on from one location. Each lamp is in parallel to the power supply.

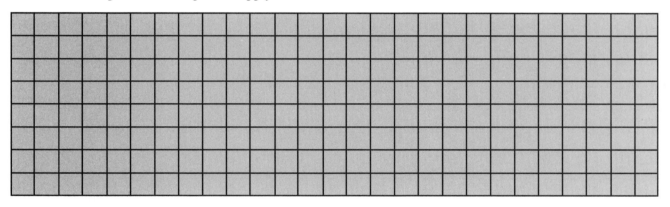

Problem 2. Draw two lamps connected in parallel that can be controlled from two locations.

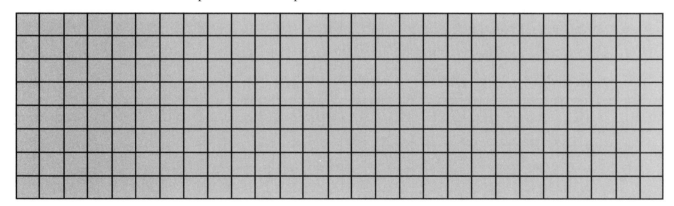

(Continued)

Problem 3. Draw three lamps connected in parallel. The lamps should be controlled by two SPST switches that are in series with each other. For the lamps to energize, both switches must be closed.

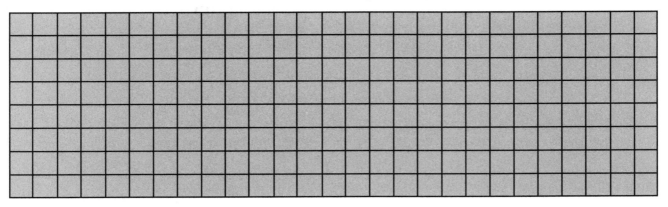

Problem 4. Draw a schematic to meet the following design conditions.

A. Must use three switches and three lamps. Be sure to label all devices.

B. Switch A must be closed before any lamp can be energized. If switch A opens, all lamps will go out.

C. Switch B will light Lamp 1, but only when Switch A is closed.

D. Switch C will light Lamp 2 and Lamp 3, but only when Switch A is closed.

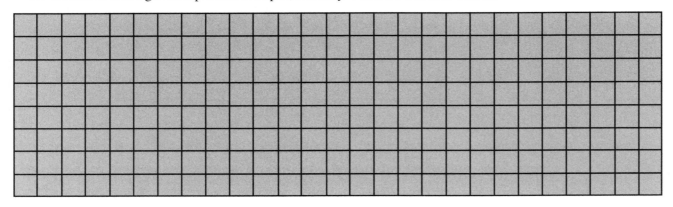

Problem 5. Draw a schematic to meet the following conditions. Label all devices.

A. Switch A will only turn the power off when actuated.

B. Switch B will light Lamp 1.

C. Switch C will light Lamp 2.

   (Hint: You must use some type of push-button switch for one location.)

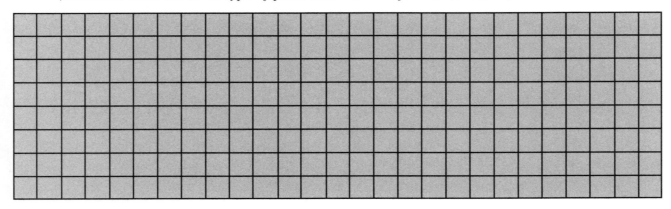

Student Activity Sheet 3-6
Schematic Design Challenge

Name_____ **Score**_____

Date_____ **Class/Period/Instructor**_____

    This is a special challenge to test your understanding of circuit design. This challenge can take some time to figure out, even for the best technician.

    Below are two buildings—Office A and Office B. A simple circuit (signaling system) must be designed for use in both offices. This signaling system is used to "ring" a person in the other office if needed. Office A has a push button to ring the bell in Office B. Office B has a push button to ring the bell in Office A.

    The problem is that only three wires were installed between the two offices. Using only the two bells, two switches, and one power supply given, you must come up with the internal wiring needed to complete the assignment. Remember, you can only use the existing three external wires between the buildings. Don't give up until you solve it.

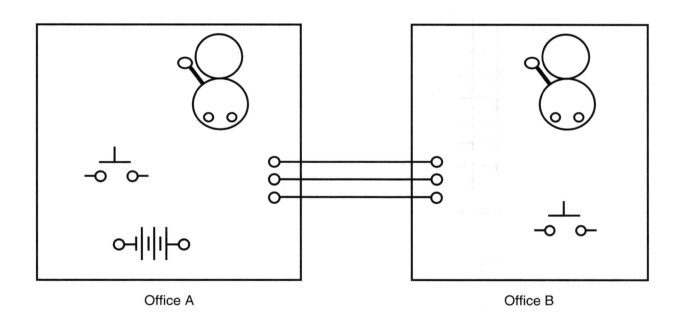

Office A   Office B

Chapter 3   Introduction to Basic Electrical Circuit Materials   **55**

Student Activity Sheet 4-1
Review

Name _____ **Score** _____

Date _____ **Class/Period/Instructor** _____

Complete the following sentences by filling in the missing word or words.

1. Work is equal to _____ multiplied by _____.

   1. _____

   _____

2. Work is expressed in units of _____.

   2. _____

3. Power is the _____ _____ of accomplishing work.

   3. _____

4. Fill in the missing parts of the following power formula.

   Power = (_____ × _____)/_____

   4. _____

   _____

   _____

5. Electrical power is measured in _____.

   5. _____

6. One watt is equal to one _____ multiplied by one _____.

   6. _____

   _____

7. One horsepower is equal to _____ watts.

   7. _____

8. Power can be found by multiplying _____ by _____, or by squaring _____ and then multiplying by the resistance.

   8. _____

   _____

   _____

9. A 10 Ω resistor connected to a 12 V power supply will produce _____ W of power lost as heat.

   9. _____

10. A 0.05 A current through a 50 Ω resistor will produce _____ W of energy as heat.

    10. _____

11. A(n) _____ is used to determine the amount of electrical energy being consumed.

    11. _____

(Continued)

12. Efficiency is determined by comparing the _____ to the _____.

12. _____

_____

13. What is the efficiency of a 1/8-hp motor that draws 5 A when connected to a source of 24 V?

13. _____

14. The twisting force of a shaft is referred to as _____.

14. _____

15. By reducing the rpm of a shaft through a system of pulleys or gears, the _____ increases.

15. _____

Student Activity Sheet 4-2
Determining Efficiency of a DC Motor

Name_____ **Score**_____

Date_____ **Class/Period/Instructor** _____

Introduction

In this experiment, you will connect a dc generator to an incandescent lamp load and determine the efficiency of the total electrical system. Most generators use a prime mover to drive the generator system such as steam, water, or combustible engine. The prime mover in this experiment will be a dc motor. You will measure the input power (current and voltage) and compare it to the output power (current and voltage). By applying the efficiency formula, you will be able to compute the efficiency of the motor generator set.

Materials and Equipment

(2)—dc motors, 9–18 volt

(1)—dc voltmeter

(1)—dc ammeter

(3)—incandescent lamps, 12 volt

(1)—12 volt dc power supply

(2)—SPST switches

(1)—breadboard

(2)—conduit straps

(1)—4″ #14 conductor

(1)—18″ #22 copper conductor

(1)—plywood base, 12″ × 24″

Procedure

Step 1. Gather all materials required for this activity.

Step 2. Couple the shafts on the dc motors using the insulation from a #14 conductor or smaller. Secure the motors to the plywood base with two 1″ conduit straps. Solder wire leads to each motor using #22 copper wire. Leads should be long enough to reach the breadboard.

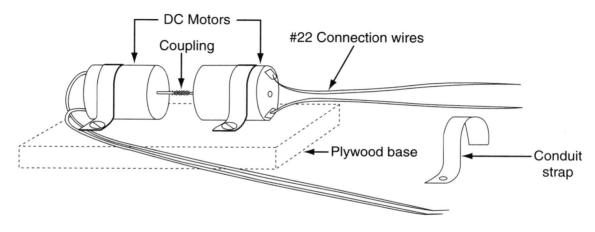

Chapter 4   Energy

(Continued)

59

Step 3. Assemble the rest of the electrical circuit on the breadboard as indicated in the following schematic-drawing.

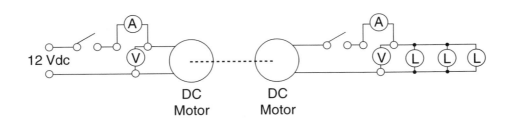

Step 4. Have your instructor check your project before energizing.

Step 5. Close the SPST switch supplying power to the dc motor. Adjust the power supply voltage to 12 Vdc. The motor should be rotating at this point. If it is not, ask your instructor for assistance.

Step 6. Close the SPST switch connecting the generator to the lamps.

Step 7. Measure the motor input voltage and current to the motor. Measure the generator output voltage and current to the lamps. Record the values below:

Motor input voltage = _____ volts

Motor input current = _____ amps

Generator output voltage = _____ volts

Generator output current = _____ amps

Step 8. Using the formula below, calculate the efficiency of the motor/generator set you have constructed.

$$\frac{\text{output volts} \times \text{amps}}{\text{input volts} \times \text{amps}} \times 100 = \text{percent efficiency}$$

_____ × 100 = _____ % efficiency

Step 9. Conduct your own experiment to determine if efficiency rating goes up or down as the load changes.

Question 1. Is efficiency affected by load? If so, how? _____

_____

An electric car is designed to run on batteries. It is equipped with a dc generator on each of its four wheels. As the electric car is driven, the four generators charge the batteries. The designer says you will create sufficient energy from the four generators to recharge the batteries. This design sounds too good to be true. Will it work or not? Explain, in detail, your support for or against the design.

_____

_____

_____

_____

_____

_____

_____

(Continued)

_____

_____

_____

_____

_____

Step 10.   Clear your work area. Properly store equipment and supplies.

Reading a Watt-Hour Meter

Name_____ **Score**_____

Date_____ **Class/Period/Instructor** _____

Interpret the watt-hour meter readings below, and then write the meter reading in the space provided.

1. Meter reading equals _____ watt-hours.

2. Meter reading equals _____ watt-hours.

3. Meter reading equals _____ watt-hours.

4. Meter reading equals _____ watt-hours.

5. Meter reading equals _____ watt-hours.

(Continued)

6. Meter reading equals _____ watt-hours.

7. Meter reading equals _____ watt-hours.

8. Meter reading equals _____ watt-hours.

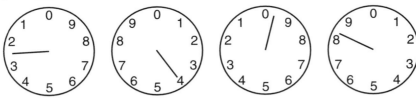

9. Meter reading equals _____ watt-hours.

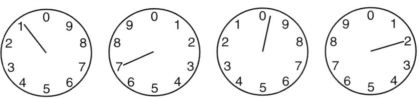

10. Meter reading equals _____ watt-hours.

11. Meter reading equals _____ watt-hours.

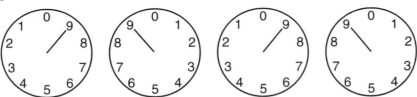

12. Meter reading equals _____ watt-hours.

Applying the Power Wheel

Name_____ **Score**_____

Date_____ **Class/Period/Instructor** _____

Technicians must be able to compute unknown values using the information provided to them. In this activity, you will use the power wheel for the calculations. Place the correct answer in the space provided. Use the area below the questions to calculate your answers.

1. A 1 kΩ resistor connected to a 6 V source will produce _____ A of current and _____ W of heat.

2. A 300 Ω, 1/4 watt resistor can be connected to a source voltage as high as _____ without exceeding its wattage rating.

3. A 200 Ω resistor conducting 0.05 A will dissipate _____ W of power.

4. A red, red, brown resistor connected to a 12 V power supply will dissipate _____ W of power.

5. The total power consumed by the circuit below is _____ W.

12 Volts   Brown, Red, Red

(Continued)

6. The maximum voltage that can be applied to a 0.5 W, 1200 Ω resistor without exceeding its wattage rating would be _____ V.

7. The maximum voltage that can be applied to a 1200 Ω, 1/4 W resistor without exceeding its wattage rating would be _____ V.

8. A 1200 Ω resistor is connected to a 12 V power supply. What is the minimum wattage resistor that can be used—1 W, 1/2 W, 1/4 W, or 1/8 W?

9. A 120 Ω resistor is connected to a 6 V power supply. What is the minimum wattage resistor that can be used—1 W, 1/2 W, 1/4 W, or 1/8 W?

10. A 200 Ω resistor is connected to a 12 V power supply. What wattage resistor should be used—1 W, 1/2 W, 1/4 W, or 1/8 W?

Sources of Electricity

Student Activity Sheet 5-1

Review

Name_____ **Score**_____

Date_____ **Class/Period/Instructor**_____

Complete the following statements by filling in the missing word or words.

1. There are six basic sources of electricity. Name them.

   _____

   _____

   _____

   _____

   _____

   _____

2. A battery is an example of producing electricity using _____ _____.　　　　2. _____

3. The solution in a battery is the _____.　　　　3. _____

4. A(n) _____ cell is often referred to as a dry cell.　　　　4. _____

5. _____ has the highest negative potential of all metals.　　　　5. _____

6. The main difference between primary cells and secondary cells is that secondary cells can be _____.　　　　6. _____

7. Specific gravity of a battery is measured using a(n) _____.　　　　7. _____

8. A(n) _____ cell is a type of battery that cannot be recharged.　　　　8. _____

9. Connecting batteries in _____ increases their voltage output.　　　　9. _____

10. Connecting batteries in _____ increases their ampacity.　　　　10. _____

11. A 20 ampere-hour battery will produce 4 amperes of current for approximately _____ hours.　　　　11. _____

12. The amount of time a battery will last before it needs recharging is indicated by its _____ rate.　　　　12. _____

13. A(n) _____ _____ can produce an electrical potential from sunlight.　　　　13. _____

14. A(n) _____ can produce an electrical potential from heat.　　　　14. _____

(Continued)

15. A meter that can measure very small amounts of current is the _____.

16. A(n) _____ can produce an electrical potential from mechanical pressure.

17. The _____ _____ produces no pollution and its only by-product is water.

18. An electrical potential is produced when an ionized gas is passed through a magnetic field. This system of electrical production is called _____.

19. An anode displays a(n) _____ polarity.

20. The cathode displays a(n) _____ polarity.

15. _____

16. _____

17. _____

18. _____

19. _____

20. _____

Simple Electrical Cell

Name _____ **Score** _____

Date _____ **Class/Period/Instructor** _____

Introduction

In this activity, you will construct a simple electrical cell consisting of a saltwater electrolyte and plates of various metals. You will take voltage readings to determine which combination creates the greatest potential of electrical energy. Record your findings in the chart below.

Materials and Equipment

Various types of metal plates:

Lead	Zinc
Copper	Aluminum
Iron	Silver

(1)—12–15 oz. tap water

(6)—tablespoons of salt

(1)—beaker

(1)—measuring spoon

(1)—multimeter or galvanometer

(2)—meter leads with alligator clips

Procedure

Step 1. Gather all materials required for this activity.

Step 2. Dissolve 4 tablespoons of salt into 8 ounces of water.

Step 3. Place the lead and the zinc plates in the saltwater electrolyte solution. Connect the meter leads to the meter. Attach the alligator clips to the plates. The setup should look similar to the following figure.

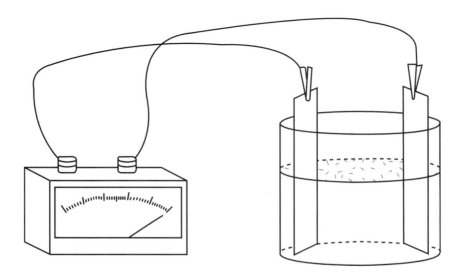

(Continued)

Step 4. Take a voltage reading of the potential created by the two dissimilar metals in the solution. Record your findings in the following chart.

	LEAD	**ZINC**	**COPPER**	**ALUMINUM**	**IRON**	**SILVER**
LEAD	V	V	V	V	V	V
ZINC	V	V	V	V	V	V
COPPER	V	V	V	V	V	V
ALUMINUM	V	V	V	V	V	V
IRON	V	V	V	V	V	V
SILVER	V	V	V	V	V	V

Step 5. Repeat Step 3 using different combinations of the metals until the chart has been completed.

Question 1. Which combination of metals developed the highest voltage potential?

_____

Question 2. Which combination developed the lowest voltage potential?

_____

Question 3. What would happen if you connected two saltwater cells in series?

_____

_____

Step 6. Rinse the metal plates and beaker in fresh water. Thoroughly dry them. Return materials to their proper storage location.

Battery Cells in Series and Parallel

Name _____ **Score** _____

Date _____ **Class/Period/Instructor** _____

Introduction

In this activity, you will be connecting D-size batteries in series and parallel to determine the effect of output voltage. You will also conduct an experiment to verify if a battery is good or bad. Lastly, you will determine the approximate internal resistance of a battery.

Materials and Equipment

(1) — lamp, 6 volt

(1) — SPST switch

(4) — D cell battery holders

(1) — multimeter

(4) — D cells (good)

(4) — D cells (bad)

Procedure

Step 1. Gather all materials required for this activity.

Step 2. Place the four good D cells in the battery holders. Connect them in series as shown in the following figure.

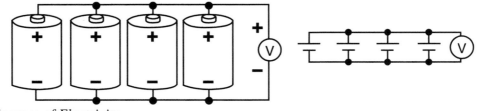

Step 3. Set the multimeter to read dc volts. Be sure to observe proper polarity when connecting the meter in the following steps.

Step 4. Take a voltage reading of one battery and record the voltage.

One cell = _____ V

Step 5. Take a voltage reading across all four batteries and record the voltage.

Four cells = _____ V

Question 1. When cells are connected in series, what happens to the voltage? _____

_____

_____

Step 6. Now connect the four good cells in parallel as shown in the following figure.

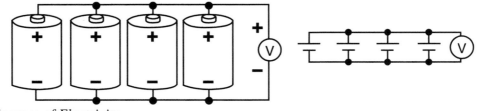

(Continued)

Step 7. Record the voltage of the total circuit.

Four cells = _____ V

Question 2. When cells are connected in parallel, what is the effect on the total voltage? _____

_____

Step 8. Connect the four good batteries in series again. This time, reverse one of the cells so that the polarity of one cell is opposite to the other three cells. Note the schematic that follows.

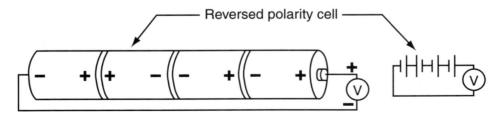

Step 9. Take a total voltage reading of this setup. Record the results.

Four cells, one reversed polarity = _____ V

Question 3. How was the total voltage affected when one of the cells was connected with reverse polarity?

_____

Question 4. What might happen if two of the four cells were connected with reversed polarity?

_____

Step 10. Prepare the meter to read dc amperage. **Note! The meter should have at least a 10-amp range.** Make sure that you connect the test leads to the proper terminals. Check with your instructor to make sure that the meter is properly connected to read amperage. (The next meter reading you take could damage the meter if the correct range has not been selected.) In general, never connect an ammeter in series with a power source, but for this experiment it is permissible. The expected values should not harm the meter. **Do not attempt this experiment with other types of power supplies.**

Step 11. Connect the meter to the one cell as in the illustration that follows. Record the amperage.

Total amperage for one cell = _____ A

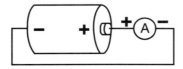

Question 5. What limited the one-cell amperage to the above reading? _____

_____

Question 6. Should this current-limiting action be considered when designing a circuit? _____

_____

Step 12. Using the procedure in Step 11, take an amperage reading of one of the bad cells. Record the value here.

Amperage of one bad cell = _____ A

Question 7. How did the amperage of the good cell compare to the bad cell? _____

_____

(Continued)

Fundamentals of Electricity and Electronics

Step 13. Connect the four bad cells, lamp, and switch in series as shown in the following figure. Leave the switch in the open position. Record the voltage.

Total voltage switch open = _____ V

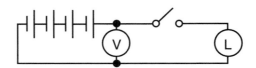

Step 14. Close the switch. Read and record the total voltage. Record the voltage.

Total voltage switch closed = _____ V

Step 15. Repeat Steps 13 and 14 using four good cells this time. Record your results.

Total voltage switch open = _____ V

Total voltage switch closed = _____ V

Question 8. What was the voltage of the good cells when compared to the voltage of the bad cells when the switch was open and when the switch was closed?

Open switch _____
_____

Closed switch _____
_____

Question 9. What would be a good procedure to follow to determine if a battery was in good or bad condition?

Explain. _____
_____
_____
_____
_____
_____
_____
_____
_____
_____

Step 16. Clear your work area. Properly store equipment and supplies.

Student Activity Sheet 5-4
Exploring a Piezo Cell

Name_____ **Score**_____

Date_____ **Class/Period/Instructor**_____

Introduction

In this lab activity, you will use a piezo cell to convert physical pressure into electricity. The primary component of a piezo cell is a crystal. When the crystal is flexed, or has a pressure applied to its surface, electrical voltage is produced. This voltage may be small, but it can be amplified to drive a powerful electrical system. Microphones made from a piezo cell crystal can drive a 5000-watt speaker system.

In the second half of this lab, you will produce sound by applying a small ac signal to the piezo cell speaker. When an electrical energy source is applied to the piezo cell, it will distort. If the energy is applied to the crystal in short pulses, it will vibrate. These vibrations can easily produce sound waves. The rate and type of waves can produce different sounds.

Materials and Equipment

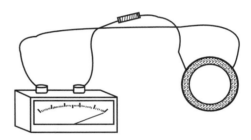

(1)—galvanometer or (1)—voltmeter

(1)—piezoelectric speaker element

(1)—signal generator

(1)—1 kΩ resistor, 1/4 W (brown, black, red)

(1)—breadboard

Procedure

Step 1. Gather all materials required for this activity.

Step 2. Connect the circuit as indicated in the following diagram. Have your instructor inspect your setup before you proceed further.

Step 3. Gently flex the piezoelectric speaker element slightly. Note the movement of the galvanometer.

Question 1. How much electrical potential does the speaker with a piezo cell unit produce?

_____

Step 4. Disconnect the speaker from the voltmeter. Now, connect the piezoelectric speaker to the signal generator as indicated below. Apply a sine wave signal, then a square wave signal, and finally a triangular wave signal. For each wave type, turn on the power and vary the frequency from 0 to 5000 Hz.

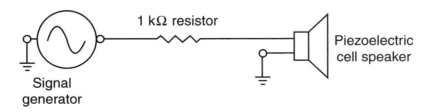

(Continued)

Question 2. Do the different wave shapes produce different sounds? _____

Question 3. What was the lowest sine wave frequency you could hear? _____

_____

Question 4. What was the highest sine wave frequency you could hear? _____

_____

Step 5.   Disconnect all the components. Clear your work area. Properly store equipment and supplies.

Student Activity Sheet 5-5
Producing Electricity Using a Photovoltaic Cell

Name_____ **Score**_____

Date_____ **Class/Period/Instructor**_____

Introduction

In this lab activity, you will use a photovoltaic cell to convert sunlight or room light into electricity.

Materials and Equipment

(1)—voltmeter

(2)—silicon solar cells

(1)—breadboard

Procedure

Step 1. Gather all materials required for the activity. Use care when handling the silicon solar cell as they are fragile and can be easily shattered.

Step 2. Connect the circuit as indicated below. Have your instructor inspect your project before you proceed.

Step 3. Read the amount of voltage being produced by the cell and record the value.

_____ V

Step 4. Connect the second solar cell in series with the first cell.

Step 5. Read and record the voltage for two cells connected in series.

_____ V

Step 6. Connect the two cells in parallel and record the amount of voltage being produced.

_____ V

Question 1. When do the cells produce the most voltage—when connected in series or in parallel?

_____

Step 7. Disconnect all the components, clean up your work area, and return all parts and materials.

Student Activity Sheet 6-1

Review

Name_____ **Score**_____

Date_____ **Class/Period/Instructor**_____

Complete the following sentences by filling in the missing word, words, or formula.

1. _____ path(s) exist for current in a series circuit.

2. In a series circuit, E_T =_____.

3. Source voltage in a series circuit is equal to the _____ _____ _____ _____ _____ _____.

4. In a series circuit, the current value is _____ throughout the circuit.

5. In a series circuit, I_T = _____.

6. Total resistance in a series circuit is equal to the _____ _____ _____ _____ _____ _____.

7. Total power consumed in a series circuit is equal to the _____ _____ _____ _____ _____ _____.

8. You can use a voltmeter to locate an open circuit when troubleshooting a simple series circuit. When the voltmeter is connected across the open in the circuit, the meter will indicate a voltage equal to the _____ _____.

9. When a resistor is shorted and a voltmeter is placed across the resistor, _____ volt(s) would be indicated.

1._____

2._____

3._____
_____

4._____

5._____

6._____
_____

7._____
_____

8._____

9._____

(Continued)

10. Draw a series circuit below using appropriate symbols. The circuit will consist of two 100 Ω resistors, one 200 Ω resistor, a SPST switch, a 12 V battery, and a fuse. Indicate the current and voltage drop value at each resistor. At the 12 V source, indicate total circuit resistance, current, and power values.

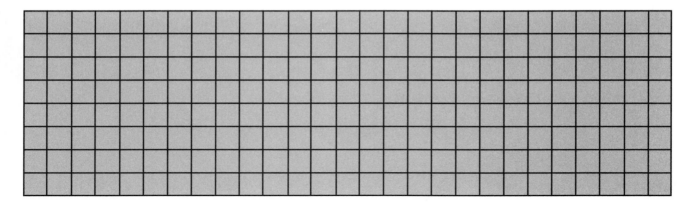

Series Circuit Practice

Name _____ **Score** _____

Date _____ **Class/Period/Instructor** _____

In series circuits you encounter in the lab, you will often know the value of some components, but not all of them. This activity will provide you with practice solving for unknown values of series circuits.

1. Fill in the missing values for the circuit below.

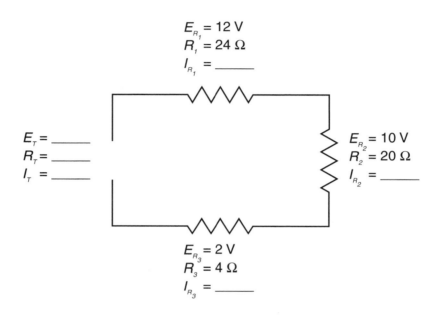

$$E_{R_1} = 12\ V$$
$$R_1 = 24\ \Omega$$
$$I_{R_1} = \underline{\hspace{1cm}}$$

$$E_T = \underline{\hspace{1cm}}$$
$$R_T = \underline{\hspace{1cm}}$$
$$I_T = \underline{\hspace{1cm}}$$

$$E_{R_2} = 10\ V$$
$$R_2 = 20\ \Omega$$
$$I_{R_2} = \underline{\hspace{1cm}}$$

$$E_{R_3} = 2\ V$$
$$R_3 = 4\ \Omega$$
$$I_{R_3} = \underline{\hspace{1cm}}$$

2. Fill in the unknown values below.

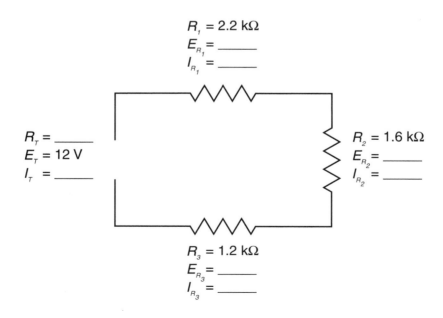

$$R_1 = 2.2\ k\Omega$$
$$E_{R_1} = \underline{\hspace{1cm}}$$
$$I_{R_1} = \underline{\hspace{1cm}}$$

$$R_T = \underline{\hspace{1cm}}$$
$$E_T = 12\ V$$
$$I_T = \underline{\hspace{1cm}}$$

$$R_2 = 1.6\ k\Omega$$
$$E_{R_2} = \underline{\hspace{1cm}}$$
$$I_{R_2} = \underline{\hspace{1cm}}$$

$$R_3 = 1.2\ k\Omega$$
$$E_{R_3} = \underline{\hspace{1cm}}$$
$$I_{R_3} = \underline{\hspace{1cm}}$$

(Continued)

3. Fill in the missing values for the circuit below.

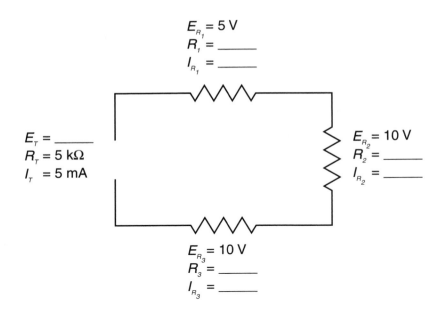

$E_{R_1} = 5\text{ V}$
$R_1 =$ _____
$I_{R_1} =$ _____

$E_T =$ _____
$R_T = 5\text{ k}\Omega$
$I_T = 5\text{ mA}$

$E_{R_2} = 10\text{ V}$
$R_2 =$ _____
$I_{R_2} =$ _____

$E_{R_3} = 10\text{ V}$
$R_3 =$ _____
$I_{R_3} =$ _____

4. Fill in the missing values below.

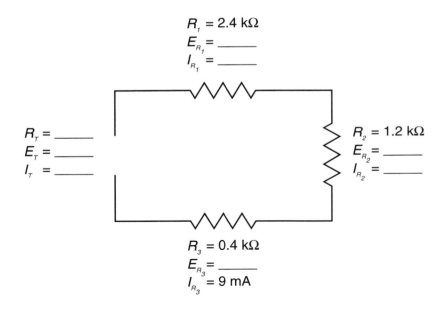

$R_1 = 2.4\text{ k}\Omega$
$E_{R_1} =$ _____
$I_{R_1} =$ _____

$R_T =$ _____
$E_T =$ _____
$I_T =$ _____

$R_2 = 1.2\text{ k}\Omega$
$E_{R_2} =$ _____
$I_{R_2} =$ _____

$R_3 = 0.4\text{ k}\Omega$
$E_{R_3} =$ _____
$I_{R_3} = 9\text{ mA}$

(Continued)

5. Fill in the missing values for the circuit below.

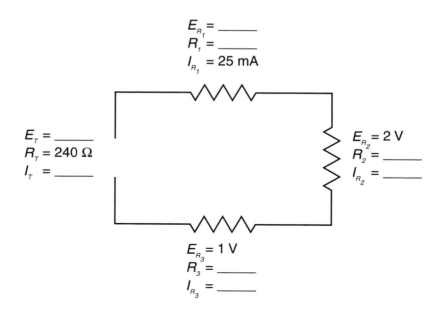

$E_{R_1} =$ _____
$R_1 =$ _____
$I_{R_1} = 25$ mA

$E_T =$ _____
$R_T = 240\ \Omega$
$I_T =$ _____

$E_{R_2} = 2$ V
$R_2 =$ _____
$I_{R_2} =$ _____

$E_{R_3} = 1$ V
$R_3 =$ _____
$I_{R_3} =$ _____

6. Fill in the missing values below.

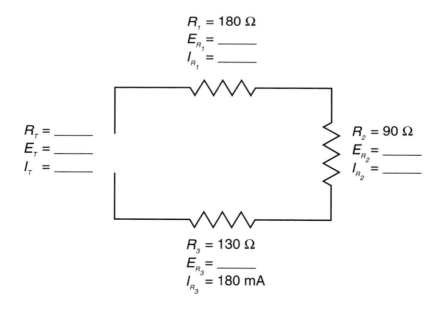

$R_1 = 180\ \Omega$
$E_{R_1} =$ _____
$I_{R_1} =$ _____

$R_T =$ _____
$E_T =$ _____
$I_T =$ _____

$R_2 = 90\ \Omega$
$E_{R_2} =$ _____
$I_{R_2} =$ _____

$R_3 = 130\ \Omega$
$E_{R_3} =$ _____
$I_{R_3} = 180$ mA

(Continued)

Chapter 6   Series Circuits

7. Fill in the missing values for the circuit below.

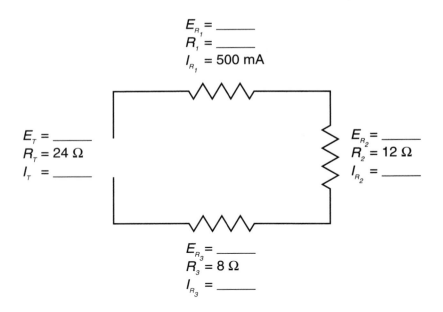

E_{R_1} = _____
R_1 = _____
I_{R_1} = 500 mA

E_T = _____
R_T = 24 Ω
I_T = _____

E_{R_2} = _____
R_2 = 12 Ω
I_{R_2} = _____

E_{R_3} = _____
R_3 = 8 Ω
I_{R_3} = _____

8. Fill in the missing values below.

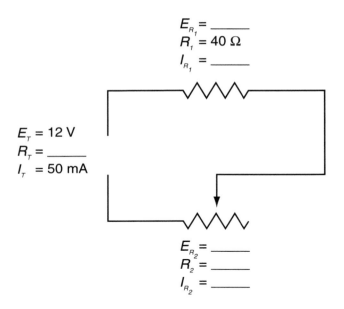

E_{R_1} = _____
R_1 = 40 Ω
I_{R_1} = _____

E_T = 12 V
R_T = _____
I_T = 50 mA

E_{R_2} = _____
R_2 = _____
I_{R_2} = _____

Student Activity Sheet 6-3
Verifying Kirchhoff's and Ohm's Laws

Name _____ **Score**_____

Date_____ **Class/Period/Instructor** _____

Introduction

In this lab activity, you will build a simple series circuit to verify Kirchhoff's law and Ohm's law. In a series circuit there is only one path for current to follow. The current value is equal throughout the circuit. The source voltage is equal to the sum of each resistor voltage drop. Total circuit resistance is equal to the sum total of the individual resistors. You will also take voltage, current, and resistance readings to become familiar with open and short circuit conditions.

Materials and Equipment

(2)—120 Ω resistors, 1/4 W (brown, red, brown)

(1)—240 Ω resistor, 1/4 W (red, yellow, brown)

(1)—12 Vdc variable power supply

(3)—short jumper wires

(1)—breadboard

(1)—SPST switch

(1)—multimeter, or (1)—ammeter, and (1)—voltmeter

Procedure

Step 1. Gather all materials required for this activity.

Step 2. Construct the following circuit. Do not energize the circuit before your instructor inspects it. Install short jumpers at the test points (indicated by TP_1, TP_2, TP_3 on the schematic.) The jumpers are used for convenience; the test points will allow you to open the circuit and install the multimeter or ammeter without disturbing the components in the circuit.

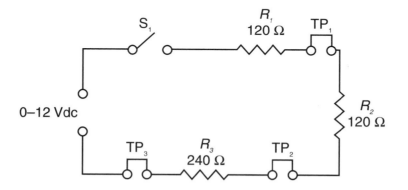

Step 3. Energize the circuit with 12 volts leaving the SPST switch in the open position. Connect the ammeter (or multimeter set to measure current) across the switch. Record the reading.

Total amperage = _____ mA

(Continued)

Step 4. Remove the jumper wire and connect the ammeter (or multimeter set to measure current) at test point 1. Close the SPST switch. Record the ammeter reading.

$$\text{Amperage value at TP}_1 = \underline{\hspace{2cm}} \text{ mA}$$

Step 5. Reinstall the jumper wire at TP_1. Remove the jumper wire at TP_2 and connect the ammeter (or multimeter set to measure current) at this location. Record the value.

$$\text{Amperage value at TP}_2 = \underline{\hspace{2cm}} \text{ mA}$$

Step 6. Reinstall the jumper wire at TP_2; then remove the jumper wire at TP_3. Connect the ammeter (or multimeter set to measure current) at TP_3. Record the value.

$$\text{Amperage value at TP}_3 = \underline{\hspace{2cm}} \text{ mA}$$

Question 1. What conclusion do you arrive at concerning current values in a series circuit?

_____

Step 7. Reinstall the jumper wire at location TP_3.

Step 8. Use the voltmeter (or multimeter set to measure voltage) to read the voltage across each resistor with the switch closed. Record the values below.

$$R_1 \text{ voltage drop} = \underline{\hspace{2cm}}$$

$$R_2 \text{ voltage drop} = \underline{\hspace{2cm}}$$

$$R_3 \text{ voltage drop} = \underline{\hspace{2cm}}$$

$$\text{Source voltage} = \underline{\hspace{2cm}}$$

Question 2. What is the main electrical characteristic of voltage in a series circuit?

_____

_____

Step 9. Connect the ammeter (or multimeter set to measure current) once again to the circuit at TP_1. Remove the jumper wire at TP_3 to simulate an open circuit condition.

Question 3. What happens to the current in a series circuit with an open circuit condition?

_____

_____

Step 10. Reinstall the jumpers at TP_1 and TP_3. Disconnect the ammeter or multimeter.

Step 11. Connect the voltmeter (or multimeter set to measure voltage) across the SPST switch. Read the voltage drop across the switch while the switch is open and while the switch is closed. Record your findings below.

$$\text{Voltage drop SPST open} = \underline{\hspace{2cm}}$$

$$\text{Voltage drop SPST closed} = \underline{\hspace{2cm}}$$

Step 12. Leave the voltmeter (or multimeter set to measure voltage) connected across the switch and remove TP_2 jumper wire. Repeat Step 11 and record your findings below.

$$\text{Voltage drop SPST open} = \underline{\hspace{2cm}}$$

$$\text{Voltage drop SPST closed} = \underline{\hspace{2cm}}$$

Question 4. Can an open circuit condition be verified at the SPST switch location? Why or why not?

_____

_____

_____

Question 5. A kitchen ceiling light is connected to a single-pole switch from a 120 volt source. If the lamp filament is burned out, what voltage reading would be present at the switch location? Explain.

_____

_____

_____

_____

Question 6. If the circuit in Question 5 is working properly, what voltage reading should you expect at the switch when open and when closed? Explain.

_____

_____

_____

_____

Step 13.   Use the voltage drop readings and the current readings taken previously to determine the power consumed at each location. Record the values below. Remember that the current through the circuit is equal in all parts of the series circuit, and the voltage is equal to the voltage drop recorded at each location. Power = Volts × Amps.

$$\text{Power at } R_1 = \underline{\hspace{2cm}} \text{ W}$$

$$\text{Power at } R_2 = \underline{\hspace{2cm}} \text{ W}$$

$$\text{Power at } R_3 = \underline{\hspace{2cm}} \text{ W}$$

Question 7. When constructing prototype circuits, how would you determine the wattage size of a resistor for the circuit? Explain.

_____

_____

_____

Question 8. Is the wattage size of a resistor equal to the amount of power it consumes or the amount of heat it can safely dissipate?

_____

_____

_____

Step 14.   Complete the chart below using the readings you have taken throughout this activity.

	Resistance Value	**Current Value**	**Voltage Drop Value**	**Power Consumed**
R_1				
R_2				
R_3				
Totals				

(Continued)

Step 15. Without using specific values, complete the formulas below using the chart as an aid.

$$R_T = \underline{\hspace{3cm}}$$

$$I_T = \underline{\hspace{3cm}}$$

$$V_T = \underline{\hspace{3cm}}$$

$$P_T = \underline{\hspace{3cm}}$$

Question 9. How did the voltage drop of R_3 compare to the voltage drop of R_1?

_____

_____

_____

Step 16. Place a jumper across R_2 to simulate a short circuit condition.

Step 17. Take a voltage reading across R_1 and R_3 and an ammeter reading at TP_1. Record the information.

Voltage drop at R_1 _____ V

Voltage drop at R_3 _____ V

Current reading at TP_1 _____ A

Question 10. Compare the readings taken in Step 15 under short circuit conditions. What happens in a typical circuit that has one component short circuited when connected in series with other components? Compare the readings and write a conclusion below.

_____

_____

_____

Step 18. Clear your work area. Properly store equipment and supplies.

Variable Resistance and Voltage
Kirchhoff's and Ohm's Laws

Name _____ **Score** _____

Date _____ **Class/Period/Instructor** _____

Introduction

In this lab activity, you will explore the relationship of a varying resistance in a series circuit and its effects on current and voltage drops. You will also observe the relationship of a variable voltage and current in a series circuit.

Materials and Equipment

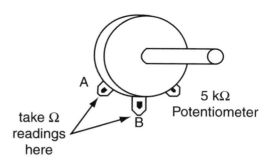

(1)—0–12 Vdc power supply

(1)—5 kΩ potentiometer

(1)—4.7 kΩ resistor, 1/4 W (orange, violet, red)

(1)—short jumper wire

(1)—SPST switch

(1)—multimeter, or (1)—voltmeter,

    (1)—ohmmeter, and (1)—ammeter

Procedure

Step 1. Gather all materials required for this activity.

Step 2. Use the ohmmeter (or multimeter set to measure resistance) to measure the resistance value of the potentiometer. Determine which terminals display the highest resistance value in relationship to a full clockwise (CW) and counterclockwise (CCW) position of the potentiometer dial or knob. During the lab activity, you will be required to adjust the potentiometer to its full resistance value and lowest resistance value. Once you have determined these values, record the information below for later reference.

Full clockwise position, terminals A to B = _____ Ω

Full counterclockwise position, terminals A to B = _____ Ω

Step 3. Connect the circuit below. Have your instructor check the circuit before energizing it. Be sure to insert a jumper wire at TP₁ as a test point for your ammeter.

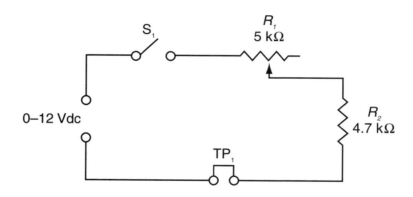

(Continued)

Step 4. Adjust the potentiometer to its full resistance value.

Step 5. Energize the circuit with 12 Vdc and close the switch. Record the voltage drop readings across the potentiometer (R_1) and the 4.7 kΩ resistor (R_2).

Voltage drop across R_1 = _____

Voltage drop across R_2 = _____

Question 1. Are the voltage drops approximately equal? _____ (The value drops should be approximately equal. If not, try adjusting the potentiometer to the opposite [CW or CCW] position.)

Step 6. Remove the jumper at TP_1. Connect the ammeter (or multimeter set to measure current) and read the current value. Record the value here.

Current value = _____ mA

Step 7. Adjust the potentiometer through its full range of resistance values. Record the current reading for full resistance and lowest resistance.

Current value, full resistance = _____ mA

Current value, lowest resistance = _____ mA

Question 2. Describe the relationship of resistance and current in a series circuit.

_____

_____

_____

Step 8. Adjust the potentiometer through its full resistance value range. Read and record the voltage drop values across R_1 and R_2.

Full resistance value R_1 voltage drop = _____; R_2 voltage drop = _____

Lowest resistance value R_1 voltage drop = _____; R_2 voltage drop = _____

Question 3. Describe the relationship of the voltage drops across the two resistors as R_1 is varied from full resistance to lowest resistance. When was the voltage drop across R_2 the greatest? Why?

_____

_____

_____

_____

Step 9. Insert the ammeter at TP_1 and record the current value. Adjust the variable power supply to 6 volts and record the current value again.

Current value at TP_1 with 12-volt source = _____

Current value at TP_1 with 6-volt source = _____

Question 4. Describe the relationship between lowering the value of source voltage and current in a series circuit when the total resistance remains the same.

_____

_____

_____

_____

Step 10. Clear your work area. Properly store equipment and supplies.

Parallel Circuits

Student Activity Sheet 7-1

Review

Name_____ **Score**_____

Date_____ **Class/Period/Instructor**_____

Complete the following sentences by filling in the missing word or words.

1. How many paths does a parallel circuit provide for current?   1. _____

2. In a parallel circuit, the voltage at each resistor is _____ (greater   2. _____
   than/equal to/less than) the source voltage.

3. Total resistance of a parallel circuit is always less than the _____   3. _____
   (smallest/largest) resistor.

4. What formula is used to determine the total resistance of a parallel circuit   4. _____
   composed of two equal resistors?

5. The _____ method of solving for total resistance of a parallel circuit   5. _____
   converts the resistor values to fractions before solving for total resistance.

6. When two unequal resistors are connected in parallel, the total resistance   6. _____
   is determined by the product over sum method. In using this method, the
   product of the resistor values is _____ (divided by/multiplied by/added
   to) the sum of the two resistor values.

7. Two 20-ohm resistors connected in parallel will have a total resistance   7. _____
   of _____.

8. Total power in a parallel circuit is equal to _____.   8. _____

9. Two 6-ohm resistors connected in parallel to a 12-volt source will   9. _____
   consume a total of _____ watts.

(Continued)

10. Draw a parallel circuit in the space provided. The circuit should contain three resistors in parallel—a 1200-ohm resistor, a 600-ohm resistor, and a 200-ohm resistor. The circuit should also include an SPST switch to control all current to the resistors. A 12-volt battery is the source for the circuit. After drawing the circuit, calculate the voltage drop at each resistor and expected current.

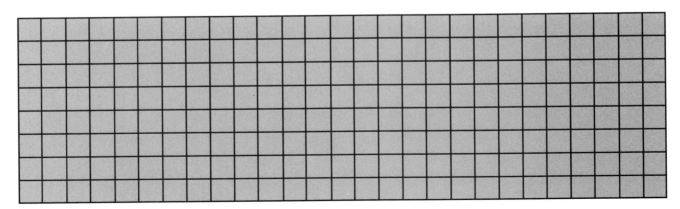

Parallel Circuit Practice

Name _____ **Score** _____

Date _____ **Class/Period/Instructor** _____

In this exercise, you will solve for the requested circuit values.

1. Find the total resistance of the three resistors connected in parallel. $R_T =$ _____ Ω

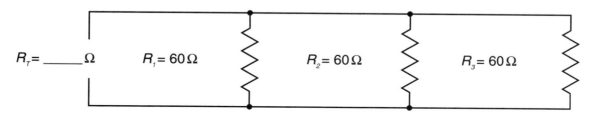

2. Find the total resistance for the two unequal resistors connected in parallel. $R_T =$ _____ Ω

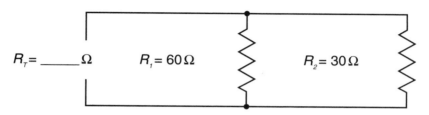

3. Find the total resistance of the three unequal resistors connected in parallel. $R_T =$ _____ Ω

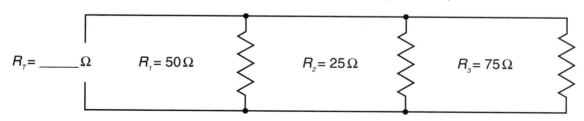

4. Calculate the missing values in the following circuit.

$R_T =$ _____ Ω $R_1 = 20\,\Omega$ $R_2 = 40\,\Omega$ $R_3 = 80\,\Omega$

$E_T = 120$ V $E_{R_1} =$ _____ V $E_{R_2} =$ _____ V $E_{R_3} =$ _____ V

$I_T =$ _____ A $I_{R_1} =$ _____ A $I_{R_2} =$ _____ A $I_{R_3} =$ _____ A

(Continued)

5. Calculate the missing values in the following circuit.

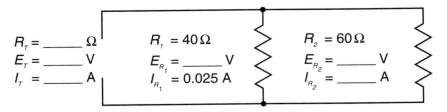

$R_T =$ _____ Ω
$E_T =$ _____ V
$I_T =$ _____ A

$R_1 = 40\,\Omega$
$E_{R_1} =$ _____ V
$I_{R_1} = 0.025$ A

$R_2 = 60\,\Omega$
$E_{R_2} =$ _____ V
$I_{R_2} =$ _____ A

6. Calculate the missing values in the following circuit.

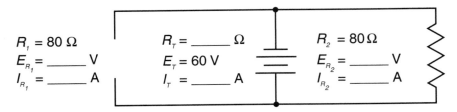

$R_1 = 80\,\Omega$
$E_{R_1} =$ _____ V
$I_{R_1} =$ _____ A

$R_T =$ _____ Ω
$E_T = 60$ V
$I_T =$ _____ A

$R_2 = 80\,\Omega$
$E_{R_2} =$ _____ V
$I_{R_2} =$ _____ A

7. Calculate the missing values in the following circuit.

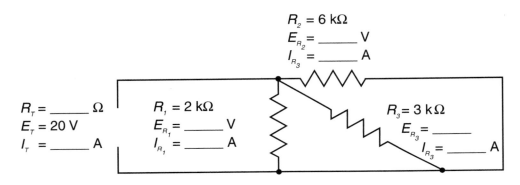

$R_2 = 6\ k\Omega$
$E_{R_2} =$ _____ V
$I_{R_3} =$ _____ A

$R_T =$ _____ Ω
$E_T = 20$ V
$I_T =$ _____ A

$R_1 = 2\ k\Omega$
$E_{R_1} =$ _____ V
$I_{R_1} =$ _____ A

$R_3 = 3\ k\Omega$
$E_{R_3} =$ _____
$I_{R_3} =$ _____ A

8. Each resistor in the following circuit is equal to 12 ohms. The source voltage is equal to 12 volts. What is the total resistance for the circuit and what is the total current?

$$R_T = \_\_\_\_\_\ \Omega;\ I_T = \_\_\_\_\_\ A$$

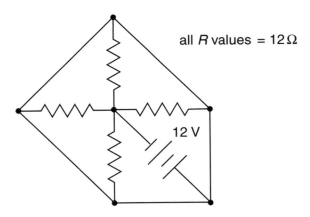

all R values = $12\,\Omega$

12 V

Verifying Kirchhoff's and Ohm's Law

Name_____ Score_____

Date_____ Class/Period/Instructor _____

Introduction

In this lab activity, you will verify Kirchhoff's law and Ohm's law by building a simple parallel circuit. In a parallel circuit, there is more than one path for electron flow. The current values through the individual resistors will vary according to their individual resistance values. The source voltage will be equally applied to each resistor. Total circuit resistance will be equal to source voltage divided by total current.

You will also take voltage, current, and resistance readings to become familiar with open and short circuit conditions. Learning the characteristics of parallel circuits will assist you tremendously when it is time for you to troubleshoot equipment or your own lab circuits.

Materials and Equipment

    (2)—120 Ω resistors, 1/4 W (brown, red, brown)

    (1)—240 Ω resistor, 1/4 W (red, yellow, brown)

    (1)—0–12 Vdc variable power supply

    (4)—short jumper wires

    (1)—breadboard

    (1)—SPST switch

    (1)—multimeter, or (1)—ammeter, (1)—voltmeter, and (1)—ohmmeter

Procedure

Step 1. Gather all materials required for this activity.

Step 2. Construct the circuit that follows. Have your instructor inspect the circuit before energizing it. Install short jumpers at the test points (TP$_1$, TP$_2$, TP$_3$, TP$_4$) indicated on the schematic. The test points will allow you to open the circuit and install the ammeter without disturbing the components in the circuit.

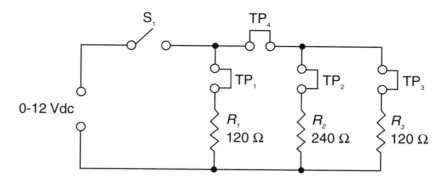

Step 3. Energize the circuit with 12 volts leaving the SPST switch in the open position. Connect the ammeter across the switch to provide you with the total current value for the circuit. Record the ampere reading.

Total current = _____ mA

(Continued)

Step 4. Remove the jumper wire. Connect the ammeter at TP₁. Close the switch and record the ammeter reading.

Current value at TP₁ = _____ mA

Step 5. Reinstall the jumper wire at TP₁. Remove the jumper wire at TP₂ and connect the ammeter at this location. Record the current value.

Current value at TP₂ = _____ mA

Step 6. Reinstall the jumper wire at TP₂ and then remove the jumper wire at TP₃. Connect the ammeter at TP₃ and record the current value.

Current value at TP₃ = _____ mA

Question 1. What conclusion do you arrive at concerning current values in a parallel circuit?

_____

_____

Step 7. Reinstall the jumper wire at TP₃.

Step 8. With the switch closed, use the voltmeter to read across each resistor and the source. Record the voltage readings in the appropriate blank below.

R₁ Voltage drop = _____

R₂ Voltage drop = _____

R₃ Voltage drop = _____

Source voltage = _____

Question 2. What is the main electrical characteristic of voltage in a parallel circuit?

_____

_____

Step 9. Connect the ammeter once again to the circuit at TP₄. Remove the jumper wire at TP₃. The removal of the TP₃ jumper will simulate an open circuit condition.

Question 3. What happened to the current in this parallel circuit with an open circuit condition?

_____

_____

_____

Step 10. Reinstall the jumpers at TP₃ and TP₄. Disconnect the ammeter.

Step 11. Connect the voltmeter across the SPST switch. Read the voltage drop across the switch while the switch is open and once again while the switch is closed. Record your findings below.

Voltage drop SPST open = _____

Voltage drop SPST closed = _____

Step 12. Leave the voltmeter connected across the switch and remove the TP₂ jumper wire. Repeat Step 11 and record your findings below.

Voltage drop SPST open = _____

Voltage drop SPST closed = _____

(Continued)

Question 4. Can this example of an open parallel circuit condition be verified at the SPST switch location?

Explain. _____

_____

_____

_____

Step 13. Use the voltage drop readings and the current readings taken previously to determine the power consumed at each location. Record those values in the spaces provided. The voltage through the circuit is equal in all parts of the parallel circuit, and the source voltage is equal to the voltage drop recorded at each location. Power = Voltage × Current.

Power at R_1 = _____ W

Power at R_2 = _____ W

Power at R_3 = _____ W

Step 14. Using the readings you have taken throughout this activity, complete the following chart.

	Resistance Value	Current Value	Voltage Value	Power Consumed
R_1				
R_2				
R_3				
Totals				

Step 15. Solve for the requested values using the chart as an aid.

R_T = _____ V_T = _____

I_T = _____ P_T = _____

Question 5. How did the current value through R_2 compare to the current value through R_1?

_____

_____

_____

_____

Step 16. **Disconnect all power from the circuit.** During this procedure, you will be using an ohmmeter. If the circuit is energized, you may damage the meter. With the circuit disconnected from the power source, read the total circuit resistance using the ohmmeter. Record the total resistance from the ohmmeter reading.

Total circuit resistance indicated by meter = _____ ohms

Step 17. Compare the total resistance reading to the calculated reading in the chart above. They should be similar. If they are not, have your instructor check your procedure.

(Continued)

Step 18. Be sure all test point jumpers are installed in the circuit. Connect the ohmmeter across R_I — one of the 120 Ω resistors. Record your finding.

Resistance R_I = _____ ohms

The resistance reading will not match the R_I resistor when all test point jumpers are left in position. When taking a resistance reading of a component that is connected in parallel with another component, you will get "backfeed reading." This means that the ohmmeter not only reads through the R_I resistor but also through all other components connected in parallel with R_I.

Step 19. Remove TP_1, and again, take a resistance reading across R_I with the ohmmeter. Record the value with TP_1 jumper removed.

R_I resistance = _____ ohms

This time the reading should be approximately equal to the color code value. It is important to understand that when taking resistance readings of components connected in parallel, at least one side of the component should be disconnected from the rest of the circuit. When circuits are soldered in place on a circuit board, it may not be convenient. You may have to rely on a total current reading and compare it to a calculated current reading to give you an indication of whether or not the components are bad. Electrical theory and simple electrical readings are the key combination to effective troubleshooting techniques.

Step 20. Clear your work area. Properly store equipment and supplies.

Combination Circuits (Series-Parallel)

Student Activity Sheet 8-1

Combination Circuit Practice

Name_____ **Score**_____

Date_____ **Class/Period/Instructor**_____

1. Calculate the missing values in the following circuit.

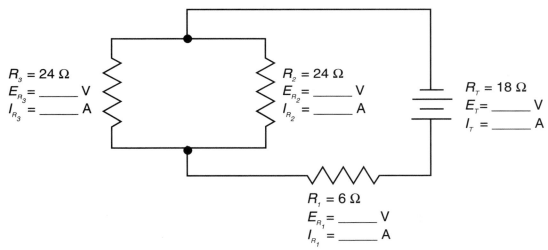

$R_3 = 24\ \Omega$
$E_{R_3} = $ _____ V
$I_{R_3} = $ _____ A

$R_2 = 24\ \Omega$
$E_{R_2} = $ _____ V
$I_{R_2} = $ _____ A

$R_T = 18\ \Omega$
$E_T = $ _____ V
$I_T = $ _____ A

$R_1 = 6\ \Omega$
$E_{R_1} = $ _____ V
$I_{R_1} = $ _____ A

2. Calculate the missing values in the following circuit.

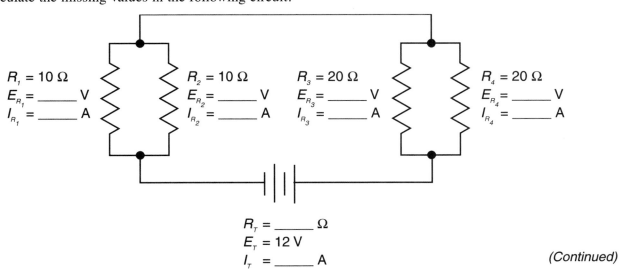

$R_1 = 10\ \Omega$
$E_{R_1} = $ _____ V
$I_{R_1} = $ _____ A

$R_2 = 10\ \Omega$
$E_{R_2} = $ _____ V
$I_{R_2} = $ _____ A

$R_3 = 20\ \Omega$
$E_{R_3} = $ _____ V
$I_{R_3} = $ _____ A

$R_4 = 20\ \Omega$
$E_{R_4} = $ _____ V
$I_{R_4} = $ _____ A

$R_T = $ _____ Ω
$E_T = 12\ V$
$I_T = $ _____ A

(Continued)

3. Calculate the missing values in the following circuit.

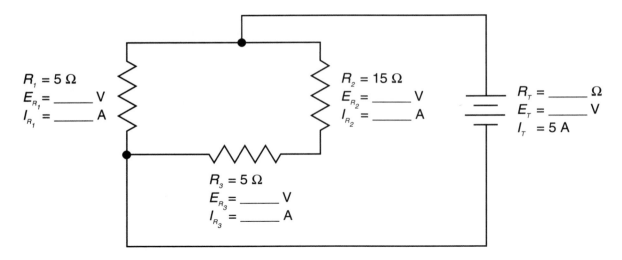

$R_1 = 5\ \Omega$
$E_{R_1} = \_\_\_\_\_$ V
$I_{R_1} = \_\_\_\_\_$ A

$R_2 = 15\ \Omega$
$E_{R_2} = \_\_\_\_\_$ V
$I_{R_2} = \_\_\_\_\_$ A

$R_T = \_\_\_\_\_\ \Omega$
$E_T = \_\_\_\_\_$ V
$I_T = 5$ A

$R_3 = 5\ \Omega$
$E_{R_3} = \_\_\_\_\_$ V
$I_{R_3} = \_\_\_\_\_$ A

4. Calculate the missing values in the following circuit.

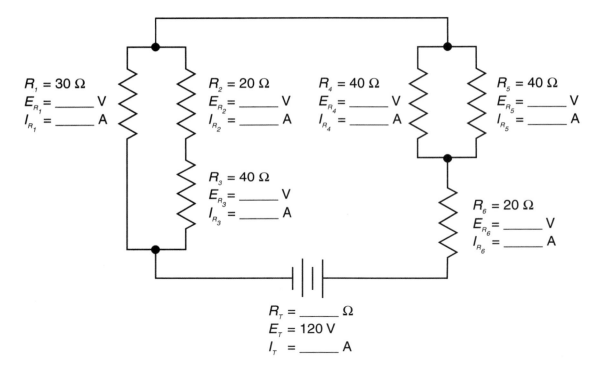

$R_1 = 30\ \Omega$
$E_{R_1} = \_\_\_\_\_$ V
$I_{R_1} = \_\_\_\_\_$ A

$R_2 = 20\ \Omega$
$E_{R_2} = \_\_\_\_\_$ V
$I_{R_2} = \_\_\_\_\_$ A

$R_4 = 40\ \Omega$
$E_{R_4} = \_\_\_\_\_$ V
$I_{R_4} = \_\_\_\_\_$ A

$R_5 = 40\ \Omega$
$E_{R_5} = \_\_\_\_\_$ V
$I_{R_5} = \_\_\_\_\_$ A

$R_3 = 40\ \Omega$
$E_{R_3} = \_\_\_\_\_$ V
$I_{R_3} = \_\_\_\_\_$ A

$R_6 = 20\ \Omega$
$E_{R_6} = \_\_\_\_\_$ V
$I_{R_6} = \_\_\_\_\_$ A

$R_T = \_\_\_\_\_\ \Omega$
$E_T = 120$ V
$I_T = \_\_\_\_\_$ A

(Continued)

5. Calculate the missing values in the following circuit.

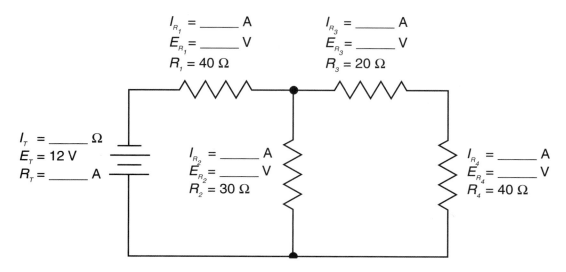

I_{R_1} = _____ A
E_{R_1} = _____ V
$R_1 = 40\ \Omega$

I_{R_3} = _____ A
E_{R_3} = _____ V
$R_3 = 20\ \Omega$

I_T = _____ Ω
$E_T = 12$ V
R_T = _____ A

I_{R_2} = _____ A
E_{R_2} = _____ V
$R_2 = 30\ \Omega$

I_{R_4} = _____ A
E_{R_4} = _____ V
$R_4 = 40\ \Omega$

6. Calculate the missing values in the following circuit. Hint: Remember Kirchhoff's current law.

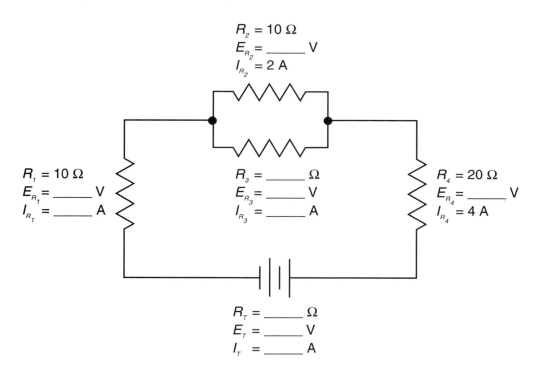

$R_2 = 10\ \Omega$
E_{R_2} = _____ V
$I_{R_2} = 2$ A

$R_1 = 10\ \Omega$
E_{R_1} = _____ V
I_{R_1} = _____ A

R_3 = _____ Ω
E_{R_3} = _____ V
I_{R_3} = _____ A

$R_4 = 20\ \Omega$
E_{R_4} = _____ V
$I_{R_4} = 4$ A

R_T = _____ Ω
E_T = _____ V
I_T = _____ A

7. Calculate the missing values in the following circuit.

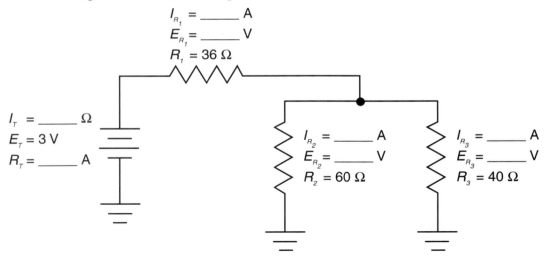

$$I_{R_1} = \underline{\hspace{1cm}} \text{ A}$$
$$E_{R_1} = \underline{\hspace{1cm}} \text{ V}$$
$$R_1 = 36 \ \Omega$$

$$I_T = \underline{\hspace{1cm}} \ \Omega$$
$$E_T = 3 \text{ V}$$
$$R_T = \underline{\hspace{1cm}} \text{ A}$$

$$I_{R_2} = \underline{\hspace{1cm}} \text{ A}$$
$$E_{R_2} = \underline{\hspace{1cm}} \text{ V}$$
$$R_2 = 60 \ \Omega$$

$$I_{R_3} = \underline{\hspace{1cm}} \text{ A}$$
$$E_{R_3} = \underline{\hspace{1cm}} \text{ V}$$
$$R_3 = 40 \ \Omega$$

8. Calculate the missing values in the following circuit.

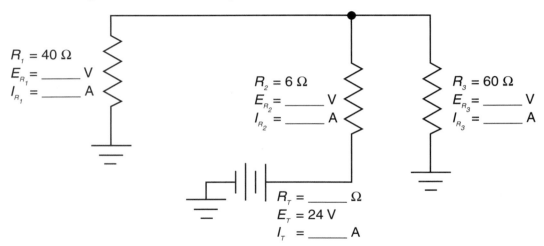

$$R_1 = 40 \ \Omega$$
$$E_{R_1} = \underline{\hspace{1cm}} \text{ V}$$
$$I_{R_1} = \underline{\hspace{1cm}} \text{ A}$$

$$R_2 = 6 \ \Omega$$
$$E_{R_2} = \underline{\hspace{1cm}} \text{ V}$$
$$I_{R_2} = \underline{\hspace{1cm}} \text{ A}$$

$$R_3 = 60 \ \Omega$$
$$E_{R_3} = \underline{\hspace{1cm}} \text{ V}$$
$$I_{R_3} = \underline{\hspace{1cm}} \text{ A}$$

$$R_T = \underline{\hspace{1cm}} \ \Omega$$
$$E_T = 24 \text{ V}$$
$$I_T = \underline{\hspace{1cm}} \text{ A}$$

9. Calculate the missing values in the following circuit.

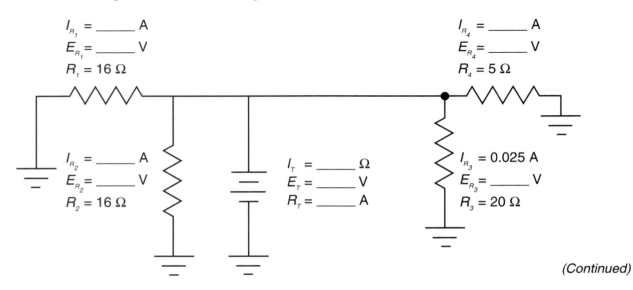

$$I_{R_1} = \underline{\hspace{1cm}} \text{ A}$$
$$E_{R_1} = \underline{\hspace{1cm}} \text{ V}$$
$$R_1 = 16 \ \Omega$$

$$I_{R_4} = \underline{\hspace{1cm}} \text{ A}$$
$$E_{R_4} = \underline{\hspace{1cm}} \text{ V}$$
$$R_4 = 5 \ \Omega$$

$$I_{R_2} = \underline{\hspace{1cm}} \text{ A}$$
$$E_{R_2} = \underline{\hspace{1cm}} \text{ V}$$
$$R_2 = 16 \ \Omega$$

$$I_T = \underline{\hspace{1cm}} \ \Omega$$
$$E_T = \underline{\hspace{1cm}} \text{ V}$$
$$R_T = \underline{\hspace{1cm}} \text{ A}$$

$$I_{R_3} = 0.025 \text{ A}$$
$$E_{R_3} = \underline{\hspace{1cm}} \text{ V}$$
$$R_3 = 20 \ \Omega$$

(Continued)

Basic Circuits

10. Calculate the missing values in the following circuit. Each resistor has a value of 10 ohms.

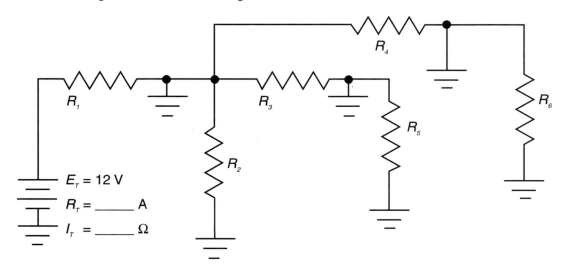

$E_T = 12$ V

$R_T =$ _____ A

$I_T =$ _____ Ω

Series-Parallel Circuit Activity
Resistance, Voltage, and Current

Name_____ **Score**_____

Date_____ **Class/Period/Instructor**_____

Introduction

In this activity, you will explore the relationship of voltage, current, and resistance in a series-parallel circuit. As the name implies, a series-parallel circuit is a combination of the two basic circuits—series and parallel. The circuit exhibits the electrical characteristics of both types of circuits. In this lab, you will verify Ohm's and Kirchhoff's laws.

Materials and Equipment

(1)—0–12 volt power supply

(2)—120 Ω resistors, 1/4 W (brown, red, brown)

(1)—240 Ω resistor, 1/4 W (red, yellow, brown)

(4)—jumper wires

(1)—SPST switch

(1)—multimeter, or (1)—voltmeter, (1)—ohmmeter, and (1)—ammeter

Procedure

Step 1. Gather all materials required for this activity.

Step 2. Set up the following circuit. Have your instructor check it before proceeding.

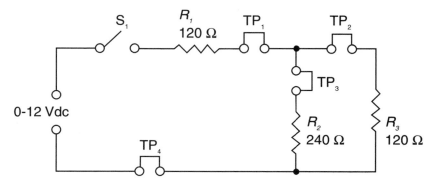

Step 3. Connect the ammeter across S_1. Turn on the power supply and adjust it for 12 volts. Take current readings at each test point location and record the values below. The current value at S_1 can be taken with the switch open, but you must close the switch to take current readings at the test point locations.

S_1 open = _____ A

TP_1 = _____ A

TP_2 = _____ A

TP_3 = _____ A

TP_4 = _____ A

(Continued)

Question 1. Which circuit component(s) carry the greatest amount of current?

_____

_____

_____

Question 2. Which circuit component(s) carry the least amount of current?

_____

_____

_____

Step 4. Using the voltmeter, take voltage drop readings at all resistor locations and S_1. Record the values below.

$$S_1 \text{ open} = \text{_____} \text{ V}$$
$$S_1 \text{ closed} = \text{_____} \text{ V}$$
$$R_1 = \text{_____} \text{ V}$$
$$R_2 = \text{_____} \text{ V}$$
$$R_3 = \text{_____} \text{ V}$$

Question 3. Describe the relationship of the voltage drops across R_2 and R_3.

_____

_____

_____

_____

Question 4. Where and when is the voltage drop equal to the total source voltage?

_____

_____

_____

Question 5. Are the voltage drops across R_2 and R_3 equal? _____

Question 6. Which components in the circuit are connected in parallel?

_____

_____

Step 6. **Disconnect the circuit from the power supply before proceeding.** You will be using the ohmmeter next and damage may result if the power supply is connected to the circuit.

Step 7. Using the test point jumpers, connect the ohmmeter across each of the resistors leaving each resistor connected in the circuit. Record the values read below.

Readings taken with circuit intact:

$$R_1 = \text{_____} \ \Omega$$
$$R_2 = \text{_____} \ \Omega$$
$$R_3 = \text{_____} \ \Omega$$

(Continued)

Step 8. Now perform the same procedure, but this time disconnect one end of each resistor and read its value with the ohmmeter. Disconnecting one end of each resistor will provide an open circuit for an accurate reading. Record your findings below.

$$R_1 = \text{\underline{\hspace{2cm}}} \ \Omega$$

$$R_2 = \text{\underline{\hspace{2cm}}} \ \Omega$$

$$R_3 = \text{\underline{\hspace{2cm}}} \ \Omega$$

Question 7. What was the cause of the variation in the readings taken in Step 7 and Step 8?

_____

Question 8. Which resistor gave the same resistance value in Step 7 and Step 8? Why?

_____

_____

_____

Step 9. Disconnect the ohmmeter. Reassemble the entire circuit and connect it back to the power supply. All resistors and test points should be reconnected in their original configuration.

Step 10. Remove TP_3 to provide an open circuit condition. Take a voltage reading across R_2 and then across TP_3. Record the values below.

$$TP_3 = \text{\underline{\hspace{2cm}}} \ \text{volts}$$

$$R_2 = \text{\underline{\hspace{2cm}}} \ \text{volts}$$

Question 9. From these readings, describe how an open conductor or component may be located in a series-parallel circuit. _____

_____

_____

_____

_____

Step 11. Clear your work area. Properly store equipment and supplies.

Student Activity Sheet 9-1
Review

Name _____ **Score** _____

Date _____ **Class/Period/Instructor** _____

Complete the following sentences by filling in the missing word or words.

1. A type of stone that was one of the first forms of natural magnets is the _____.

    1. _____

2. The angle of _____ is the angular measurement between true north and magnetic north.

    2. _____

3. All bar magnets have two poles. What are they?

    3. _____

    _____

4. The law of magnetism states that _____ poles attract each other, while _____ poles repel each other.

    4. _____

    _____

5. The invisible lines of magnetic force that surround a magnet are called the _____ _____.

    5. _____

6. As the distance between two magnets increases, the strength of the attraction between the magnets _____ (increases/decreases).

    6. _____

7. A(n) _____ _____ exists around an energized conductor.

    7. _____

8. Look at the following illustrations. According to the left hand rule for a conductor, illustrations _____ and _____ are correct.

    8. _____

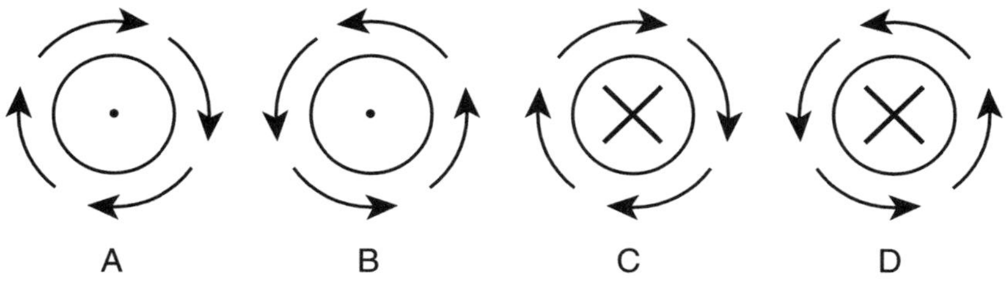

    _____

A   B   C   D

9. Resistance to magnetic flux is called _____.

    9. _____

(Continued)

10. A current-carrying conductor wound into the shape of a coil is called a(n) _____.

10. _____

11. According to the left hand rule for a coil, your thumb always points toward the _____ pole of the coil.

11. _____

12. The strength of the magnetic field of a solenoid depends on two factors. Name them.

_____

_____

13. The small amount of magnetism that remains in an electromagnet after the circuit is de-energized is called _____ magnetism.

13. _____

14. Protection from a magnetic field by using an iron object is called _____ _____.

14. _____

Conductor Magnetic Flux

Name _____ **Score** _____

Date _____ **Class/Period/Instructor** _____

Introduction

In this lab activity, you will explore the relationship of electrical current and magnetic flux surrounding an energized conductor. You will verify that a magnetic field does exist around it and how to easily increase or cancel that effect.

Materials and Equipment

(1)—compass

(1)—47 Ω resistor, 1 W (orange, violet, black)

(1)—momentary contact push-button switch

(1)—24″ #22 solid copper conductor, insulated

(1)—0–12 Vdc variable power supply

(1)—breadboard

Procedure

Step 1. Gather all materials required for this activity.

Step 2. Construct the circuit in the following diagram using the #22 conductor, resistor, and compass. **Do not energize the circuit until your instructor has inspected your work.** Align the conductor directly over the compass needle.

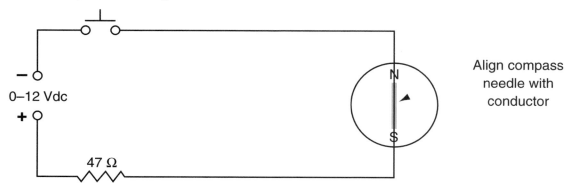

Step 3. Energize the circuit with 6 Vdc and press the push-button switch. Observe the movement of the compass needle.

Question 1. Which way did the compass needle move? Why? _____

_____

Step 4. Turn off the power supply. Reverse polarity of the power supply so the current will flow in the opposite direction over the compass needle. Re-energize the circuit and press the push-button. Observe the direction of the compass needle.

(Continued)

Question 2. What effect did reversing current direction over the compass have on the direction of compass needle deflection?_____

_____

Step 5. Reverse the polarity again so that the power supply is attached as it was in Step 1. As shown in the following illustration, shape the conductor so there will be two paths of electron flow directly over the compass needle.

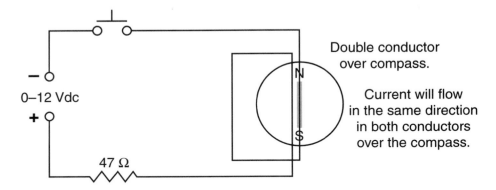

Step 6. Once again, energize the circuit with 6 Vdc.

Question 3. What effect did doubling the conductor have on the deflection of the needle?

_____

_____

Step 7. Shape the conductor to resemble the drawing that follows. This time, the circuit path will allow the current to flow in two different directions at the same time directly over the needle. Energize the circuit with the push-button switch. Observe the action of the needle.

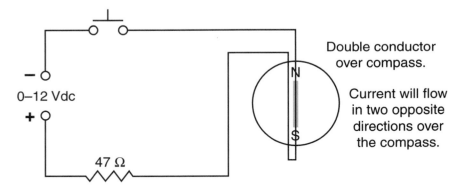

Question 4. What was the effect of the conductor flux on the compass this time? Why?

_____

_____

Step 8. Turn off the power supply, disassemble the circuit, and properly store all equipment and supplies.

Question 5. The National Electrical Code requires that current-carrying conductors in electrical wiring be run as pairs and never as a single conductor. What might be a reason for this requirement?

_____

_____

Student Activity Sheet 9-3
Electromagnet and Solenoid

Name _____ Score _____

Date _____ Class/Period/Instructor _____

Introduction

In this lab, you will build a simple solenoid and use it as an electromagnet. It will allow you to become familiar with electromagnetism in an electrical field.

Materials and Equipment

(1)—1.5 V D-cell battery with battery holder

(1)—NO momentary contact push-button switch

(1)—20d nail

(1)—10 Ω resistor, 10 W

(1)—drinking straw

(1)—compass

(1)—breadboard

(1)—120″ #30 magnet wire

— tape

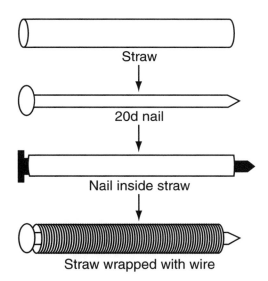

Procedure

Step 1. Gather all materials required for this activity.

Step 2. Cut the drinking straw to a length slightly shorter than the shank of the nail. Place the 20d nail inside the straw. Wind the #30 magnet wire around the straw about 120 times. Use short pieces of tape to attach the wire to the straw at both ends. You have now constructed a steel-core solenoid.

Step 3. Construct the circuit below. Have your instructor inspect the circuit before energizing it with the battery.

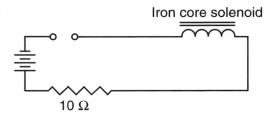

(Continued)

Step 4. Position the solenoid in an east-west direction. Energize the circuit with the push-button switch and move the compass around the energized coil. Note the direction of the compass needle as you move the compass around the coil. In the following illustration indicate the direction of the needle in each of the positions.

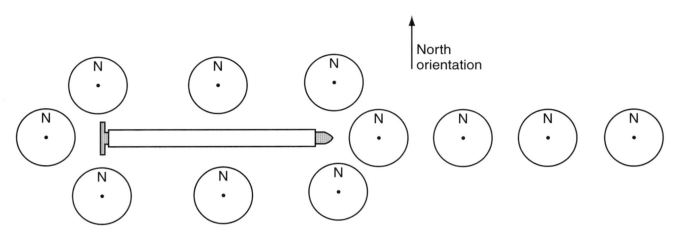

Step 5. Reverse polarity of the circuit at the power supply. Observe the effect of reversing polarity on the direction of the compass needle as you move the compass around the coil.

Question 1. What effect does polarity, or direction of current, in a coil of wire have on the magnetic field surrounding the coil? _____

_____

Question 2. Where does the magnetic field appear to be strongest? _____

_____

_____

_____

Step 6. Remove the nail from the straw. Repeat Step 4.

Question 3. Describe the effect on the magnetic strength of the coil with the nail removed from the core of the coil. _____

_____

Step 7. Remove the 10 Ω resistor from the circuit, but do *not* leave an open. Re-insert the nail into the center of the coil. Energize the circuit briefly with the push-button while repeating Step 4.

Question 4. The current is significantly increased with the 10 Ω resistor removed. What effect did increasing the current have on the magnetic field surrounding the coil and nail? _____

_____

_____

Step 8. Clear your work area. Properly store equipment and supplies.

DPDT Relay

Name_____ **Score**_____

Date_____ **Class/Period/Instructor**_____

Introduction

In this activity, you will become familiar with a common double-pole, double-throw (DPDT) relay. You will connect the relay to a 12 Vdc power source and control the energizing of the relay coil using a single-pole switch. You will use both the normally closed (NC) and the normally open (NO) contacts on the relay to energize two 12 V lamps. One lamp will be energized while the other is not. When the switch is thrown, the lamps will behave in an opposite manner.

Materials and Equipment

(1)—12 Vac power supply

(1)—12 Vdc power supply

(1)—12 Vdc DPDT relay

(1)—SPST switch

(2)—12 V lamps and lamp holders

(1)—ohmmeter

(2)—red connection wires

(2)—black connection wires

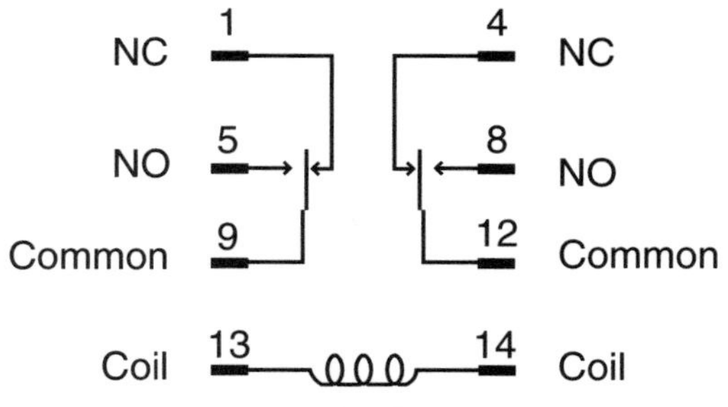

DPDT Relay

Procedure

Step 1. Gather all materials required for this activity.

Step 2. Familiarize yourself with the relay. Use the drawing shown above to identify the coil connections on the relay. The coil connection points will be 13 and 14 if you are using a Radio Shack DPDT relay (part # 275-206). If you are using another relay, have your instructor show you the correct connection points.

Step 3. Measure the resistance of the coil and record the value.

Coil resistance = _____ Ω

Step 4. Measure the contact resistance of the NO contacts and the NC contacts. Record the values here.

Terminals 1 to 9 = _____ Ω

Terminals 5 to 9 = _____ Ω

Terminals 4 to 12 = _____ Ω

Terminals 8 to 12 = _____ Ω

Question 1. On an unfamiliar relay, how might you determine the connection points for the coil circuit?

_____

_____

(Continued)

Question 2. Relay contact connection points are classified as normally open, normally closed, and common. The term "common" means that a certain connection point can complete a circuit to two or more contacts. Does the relay you are using have a common connection on the contact terminals?

_____

_____

_____

Step 5. Connect the circuit that follows. Note that you will use both an ac power supply and a dc power supply. The coil is energized with the dc supply and the lamps are energized with the ac supply. Do not cross-connect these circuits.

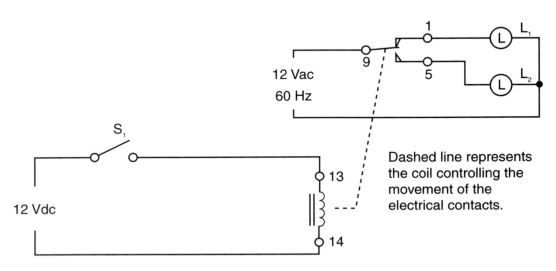

Notice the dashed line between the coil symbol and the contact symbol. This line is used to show which relay coil operates which set of the contacts. This distinction is especially important when there is more than one relay in a circuit. To make the wiring easier to follow, use red wire for the dc circuit and black wire for the ac circuit. The black and red wire should not connect to one another if the circuit is wired correctly. Have your instructor check the circuit before you energize it.

Step 7. Energize both power supplies.

Step 8. Close switch S_1 while observing the condition of the lamps L_1 and L_2. Open and close switch S_1 several times. The lamps should light alternately.

Question 3. If one lamp was used per contact, how many lamps could be controlled by S_1?_____

Question 4. Examine the relay markings. What is the maximum current and voltage rating of the contacts?

Maximum voltage of contacts = _____ V

Maximum current rating of contacts = _____ A

Step 9. Turn off both power supplies.

Step 10. Disconnect the coil circuit from the 12 Vdc supply, but do not disassemble the circuit.

Step 11. Reconnect the coil circuit of the relay to the 12 Vac power supply, the same power supply as the lamps.

Step 12. Turn on the 12 Vac power supply and close switch S_1.

Question 5. Does the relay behave the same using alternating current as it did with direct current? Explain.

_____

_____

Step 13. Clear your work area. Properly store equipment and supplies.

Motors, Generators, and Power Distribution

Emergency Lighting System

Name _____ **Score** _____

Date _____ **Class/Period/Instructor** _____

Introduction

In this activity, you will construct an emergency lighting system using a relay to switch between commercial power (alternating current) and battery backup power (direct current). In schools, offices, and many other commercial locations, there must be a way of providing emergency lighting and power when the electricity goes out.

In this project, you will simulate the commercial power source using a 12 Vdc power supply. Two 6 V batteries will be used for the emergency power supply. A relay will be used to switch between the two systems. The same design could be used for many applications other than an emergency lighting system.

In this project, you will also see another common set of schematic symbols that are used to denote the coil and contacts on a relay. These symbols are commonly used in industrial settings. (In the last activity, the schematic was illustrated using typical electronic symbols.) To be proficient in electrical systems, you should become familiar with both type of symbols.

Materials and Equipment

(2)—6 V batteries

(1)—12 Vdc power supply

(1)—SPDT relay, 12 Vdc coil

(1)—12 V lamp and lamp holder

Procedure

Step 1. Gather all materials required for this activity.

Step 2. Carefully study the following schematic and then construct the circuit. Leave the positive connection off the battery for now. Do not energize the project until your wiring has been approved by your instructor.

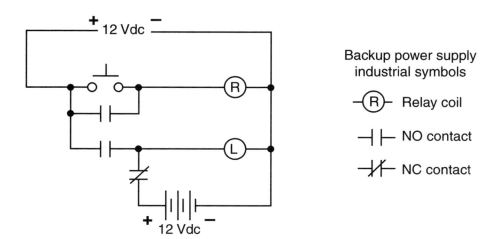

Step 3. Connect the positive lead to the 12 V battery. Then, turn on the 12 Vdc power supply.

(Continued)

Step . Push the NO push-button closed to cause the relay to energize. All NO contacts will now close and all NC contacts will open. This operation represents how an electrical system would normally be found unless there has been a power outage.

Step 5. To see if your backup circuit will operate properly in case of a power outage, simply turn off the 12 Vdc power supply. The relay should be de-energized, which will cause the lamp to be powered by the 12 Vdc battery system.

Draw the circuit below as it would appear in normal commercial power operation (not a power outage condition). The proper contact condition is essential. Would the contacts be NC or NO?

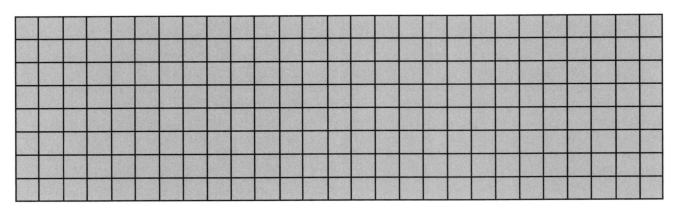

Draw the original schematic that appears after Step 2. This time, use the electronic symbols for the coil and contacts. Refer to the previous activity for relay coil connections and contact connections.

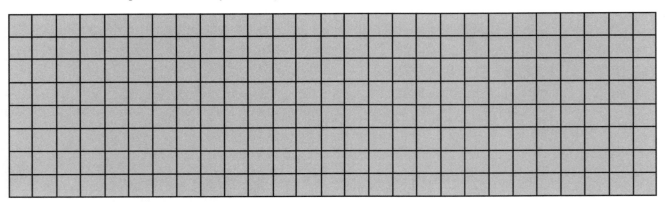

Step 6. Clear your work area. Properly store equipment and supplies.

Student Activity Sheet 10-1

Review

Name_____ **Score**_____

Date_____ **Class/Period/Instructor**_____

Complete the following sentences by filling in the missing word or words.

1. Moving a conductor through a magnetic field induces _____ in the conductor.

1._____

2. The amount of voltage produced in a generator depends upon the number of _____ lines of force and the _____ of the conductor through the magnetic field.

2._____
_____

3. _____ current flows back and forth through a conductor.

3._____

4. _____ current flows in only one direction through a conductor.

4._____

5. A generator converts _____ energy into electrical energy.

5._____

6. The revolving coil, or _____, is hung in the generator case.

6._____

7. A set of _____ are used to connect the connections of the armature to the generator load.

7._____

8. Another name for a split ring is a(n) _____.

8._____

9. List three primary reasons for generator losses.

_____

_____

_____

10. A(n) _____ _____ generator does not require a battery to energize its field poles.

10._____

11. A shunt generator has field pole windings connected in _____ (series/parallel) to the armature.

11._____

12. After the magnetizing force is removed from a field pole, a small amount of _____ _____ will remain.

12._____

(Continued)

13. A(n) _____ generator has both shunt windings and series windings.

13. _____

14. Voltage regulation of a generator is determined by comparing the _____-_____ voltage to the _____-_____ voltage.

14. _____

_____

15. The frequency of an ac voltage is measured in cycles per second, or _____.

15. _____

16. The highest point on an ac sine wave is referred to as the _____.

16. _____

17. The average voltage is equal to _____ multiplied by the peak voltage.

17. _____

18. To determine the _____ voltage, multiply 0.707 by the peak voltage.

18. _____

19. In common commercial ac generators, a rotating magnetic field is used with a stationary _____ to produce electricity.

19. _____

20. The part of a generator system that causes the rotation of the generator is a(n) _____ _____.

20. _____

21. A small dc generator mounted on the same shaft as the rotating field of a large commercial ac generator is called the _____.

21. _____

22. To connect large commercial generators in parallel, three electrical characteristics of each generator must match. Name them.

_____

_____

_____

23. Brushes must be _____ when being replaced so they will seat properly on the commutator or slip rings.

23. _____

24. A sine wave can be displayed using a(n) _____.

24. _____

25. A probe that multiplies the amplitude of a sine wave is called a(n) _____ probe.

25. _____

26. An oscilloscope should be _____ before using it to ensure accurate readings.

26. _____

27. The _____ control on an oscilloscope is used to make the image sharper.

27. _____

28. The brightness of the sine wave is adjusted by the _____ control.

28. _____

29. Why shouldn't you leave an oscilloscope unattended with an image on the screen?

_____

_____

30. The _____ should be reduced when not using the scope.

30. _____

Determining Voltage and Frequency using an Oscilloscope

Name _____ **Score** _____

Date _____ **Class/Period/Instructor** _____

In this activity, you will practice converting oscilloscope wave patterns to equivalent voltages and frequencies.

For each display, look at the wave pattern indicated on the scope grid. Pay attention to the switch settings for the scope and the probe marking. Fill in the blanks indicating voltage and frequency.

1. What is the voltage and frequency displayed on the oscilloscope grid?

Probe = ×1

Time/div. = 5 mS

Volt/div. = 0.1

Peak voltage = _____ V

Peak-to-peak voltage = _____ V

Frequency = _____ Hz

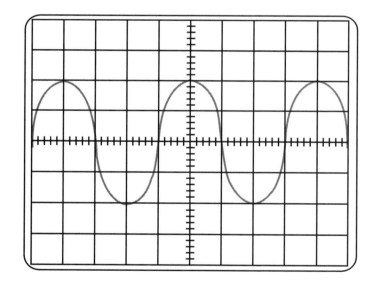

(Continued)

Chapter 10   Generators

2. What is the voltage and frequency displayed on the oscilloscope grid?

Probe = ×10

Time/div. = 20 mS

Volt/div. = 2

Peak voltage = _____ V

Peak-to-peak voltage = _____ V

Frequency = _____ Hz

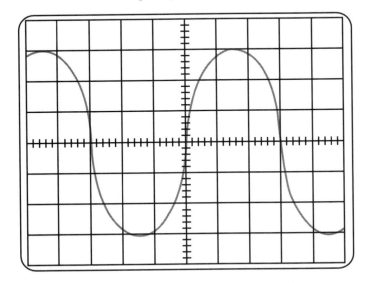

3. What is the voltage and frequency displayed on the oscilloscope grid?

Probe = ×10

Time/div. = 0.2 mS

Volt/div. = 0.02

Peak voltage = _____ V

Peak-to-peak voltage = _____ V

Frequency = _____ Hz

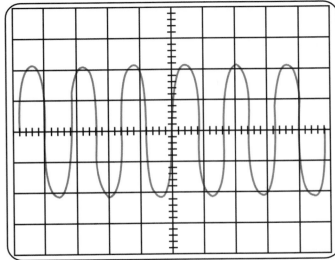

(Continued)

Motors, Generators, and Power Distribution

4. What is the voltage and frequency displayed on the oscilloscope grid?

Probe = ×1

Time/div. = 50 μS

Volt/div. = 0.1

Peak voltage = _____ V

Peak-to-peak voltage = _____ V

Frequency = _____ Hz

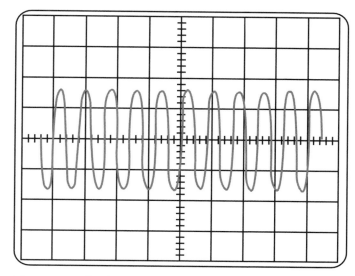

5. What is the voltage and frequency displayed on the oscilloscope grid?

Probe = ×10

Time/div. = 10 μS

Volt/div. = 5

Peak voltage = _____ V

Peak-to-peak voltage = _____ V

Frequency = _____ Hz

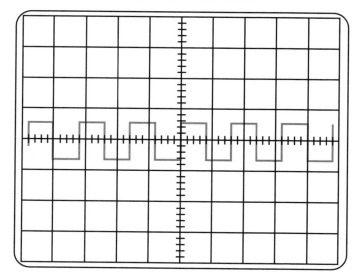

Determining Voltage and Frequency of a DC Generator

Name_____ **Score**_____

Date_____ **Class/Period/Instructor**_____

Introduction

In this lab activity, you will connect a motor generator to a resistor. One motor will serve as the prime mover and the other as a dc generator. Once this is accomplished, you will then determine the amount of voltage and frequency being produced using an oscilloscope. Then, you will raise the applied voltage to the prime mover and observe the effect on the generator output voltage and frequency.

Materials and Equipment

(2)—dc motors, 9–18 volt

(1)—0–18 volt variable dc supply

(1)—oscilloscope

(1)—dc voltmeter

(1)—1.2 kΩ resistor, 1/2 W

(1)—SPST switch

(1)—breadboard

(4)—conduit straps

(1)—plywood base (8″ × 4″)

Procedure

Step 1. Gather all materials required for this activity.

Step 2. Couple and secure the dc motors to the plywood base with two conduit straps.

Step 3. Connect the output of the generator to the 1.2 kΩ resistor as indicated in the schematic that follows.

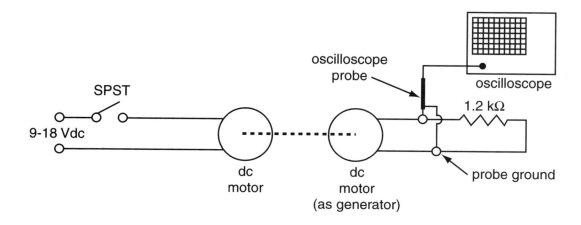

Step 4. Calibrate the oscilloscope.

Step 5. Connect the oscilloscope to the load resistor as indicated in the schematic. Be sure to connect the ground on the oscilloscope probe.

(Continued)

Step 6. Set the power supply voltage to 9 Vdc and energize the circuit by closing the switch.

Step 7. Using the dc voltmeter, read the voltage. Record the output voltage.

Output = _____ V

Step 8. Draw the wave pattern displayed on the oscilloscope.

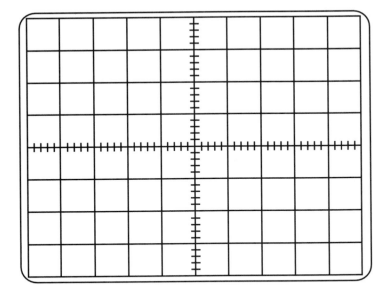

Question 1. What is the frequency of the pulsating dc voltage? (Remember, the frequency is only one complete repetition of the display wave.)_____

Frequency = _____

Step 9. Increase the power supply voltage to 18 V.

Step 10. Adjust the Time/div. until a clear wave pattern is displayed on the CRT screen. Draw the wave pattern displayed on the oscilloscope.

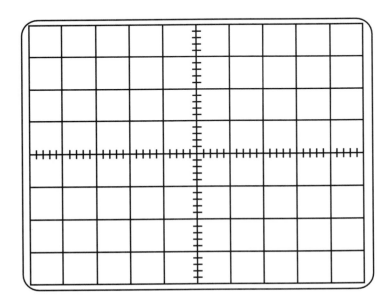

(Continued)

Motors, Generators, and Power Distribution

Question 2. How was the output voltage and frequency affected by increasing the voltage to the motor?

_____

_____

_____

_____

_____

Special Note: You will notice that the wave pattern displayed is not perfectly smooth. There are little ridges along the edge of the pattern. These small distortions are called *voltage spikes,* and are sometimes referred to as *transient voltage*. The spikes are caused by the action of the brushes making and breaking connection to the commutator segments as the generator rotates. Sometimes the voltages produced by the armature action can be quite high. If the transient voltages are very high, they can destroy sensitive electrical equipment. A filtering device built from capacitors and diodes is sometimes used to alleviate the high voltages produced. This type of filter is covered in detail later in the text.

Step 11. Clear your work area. Properly store equipment and supplies.

Student Activity Sheet 11-1

Review

Name_____ **Score**_____

Date_____ **Class/Period/Instructor**_____

Complete the following sentences by filling in the missing word or words.

1. An electric motor converts electrical power into rotating _____ power.

1. _____

2. The electromagnets used in place of permanent magnets in a dc motor are called _____ _____.

2. _____

3. The generated electrical force that is in opposition to the applied voltage is known as _____ _____.

3. _____

4. The counter emf in a dc motor will increase when _____ _____ or field strength is increased.

4. _____

5. Current in the armature circuit is high when counter emf is _____.

5. _____

6. A(n) _____ _____ condition exists when the motor cannot turn even though voltage is applied to the motor.

6. _____

7. A(n) _____ or a(n) _____ _____ is used to protect against locked rotor or severe overload conditions.

7. _____

_____

8. Sparking between the motor brushes and commutator can be reduced by using _____.

8. _____

9. Speed regulation can be calculated by subtracting the _____ speed from the _____ speed and then dividing the difference by the _____ speed. This value is then multiplied by 100 to obtain a percentage.

9. _____

_____

_____

10. The three main dc motors are _____, _____, and _____.

10. _____

_____

_____

(Continued)

11. The _____ dc motor has its field pole windings connected in parallel to the armature circuit.

11. _____

12. The _____ dc motor has its field pole windings connected in series to the armature circuit.

12. _____

13. The _____ dc motor uses both series and shunt windings.

13. _____

14. A(n) _____ compound motor has its series and shunt fields in opposition.

14. _____

15. A(n) _____ starter requires the action of a person using a lever.

15. _____

16. A(n) _____ starter is used for remote starting.

16. _____

17. A(n) _____ _____ rectifier can be used to control the speed of a dc motor.

17. _____

18. The _____ motor can run on ac and dc voltage.

18. _____

19. A(n) _____ motor has a feedback device that sends an electrical signal to the controls, providing information such as speed and position.

19. _____

20. A(n) _____ _____ sends electrical pulses to an electrical controller.

20. _____

Student Activity Sheet 11-2
Wiring Practice—Direct Current Motors

Name _____ **Score** _____

Date _____ **Class/Period/Instructor** _____

    Below are three different dc motors. Using a pencil, connect the motor leads to the source, and to each other, as indicated by the schematic to the right. This is an excellent way to practice motor connections without the danger of connecting them improperly.

  1. Connect the motor leads on the left according to the schematic on the right. Be sure to observe proper polarity.

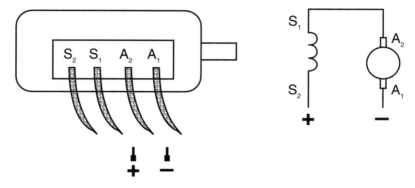

  2. Connect the motor leads on the left according to the schematic on the right. Be sure to observe proper polarity.

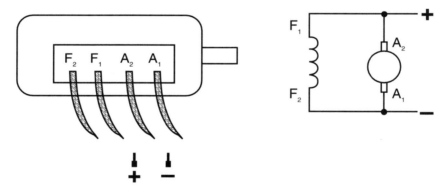

  3. Connect the motor leads on the left according to the schematic on the right. Be sure to observe proper polarity.

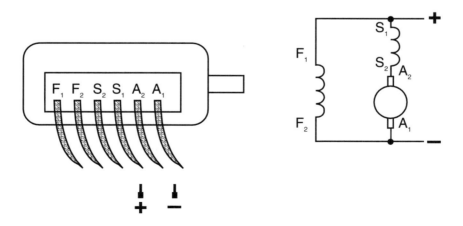

Chapter 11   DC Motors

Standard Push-Button Motor Control System

Name _____ **Score** _____

Date _____ **Class/Period/Instructor** _____

Introduction

In this lab activity, you will install and operate one of the most common control circuits used in industry to energize and de-energize motors safely. Look at the schematic while the installation and operation is explained. Two push-button switches—one normally open and the other normally closed—are used together with a relay. When the relay coil is energized by the normally open (NO) push-button switch, the relay coil is energized. When the relay coil is energized, all normally open contacts will close.

Two contacts are wired in series with the dc motor, while one contact is wired in parallel with the start push-button switch. When the contact located at the start switch closes, it will complete the circuit to the coil. This will continue to provide power to the relay coil even after the normally open push button is released. The motor is energized and will continue to run until the normally closed (NC) stop push-button switch is engaged. When the stop button opens the circuit to the coil, the coil will de-energize causing all normally open contacts to open. The dc motor will cease to run.

It is important to be familiar with the sets of standard symbols used for relays in electronic work and industrial work. The standard relay symbol varies somewhat in industrial electrical systems. A circle with a letter or abbreviation identifies the coil, while the contact is drawn with a slight variation. You should be able to identify both sets of symbols.

Materials and Equipment

(1)—12 volt dc power supply

(1)—12 volt DPDT relay

(1)—9–18 volt dc motor

(1)—normally open (NO) push-button switch

(1)—normally closed (NC) push-button switch

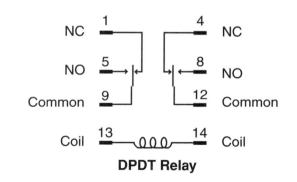

DPDT Relay

Procedure

Step 1. Gather all materials required for this activity.

Step 2. Connect the circuit according to the following schematic. Have your instructor check your work before energizing the circuit.

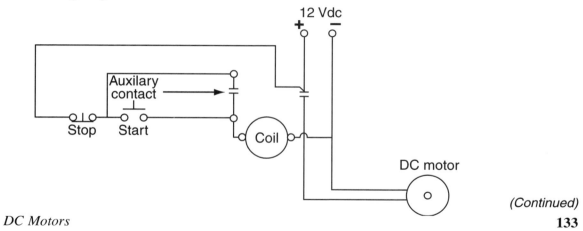

Chapter 11   DC Motors

(Continued)

Step 3. Turn on the power supply and test the circuit action. Push the NO push-button switch. The dc motor should run. Press the NC push-button switch to stop the motor. If the circuit fails to work properly, notify your instructor.

Many industrial systems utilize multiple push-button switches to allow motors to be started or stopped from more than one location. For example, a central control unit may be able to stop or start a motor, while another push-button switch directly beside the motor unit may be able to energize/de-energize the unit from that location. When more than one push-button switch is needed, all stop buttons are wired in series with each other and all start buttons are wired in parallel with each other.

Question 1. Draw the wiring diagram for a system that contains two start-stop locations in the space below.

Step 4. Clear your work area. Properly store equipment and supplies.

Student Activity Sheet 11-4
Sequential Motor Control System

Name _____ **Score** _____

Date _____ **Class/Period/Instructor** _____

Introduction

In this activity, you will connect a sequence control system using two stop-start relay systems and two dc motors. This system is commonly found in industry when the control system is designed to prevent the start of one motor before another motor is energized. For example, in a metal fabrication shop, a motor on a large milling machine is used to drive the cutting blades and another motor is used to drive the lubrication pump. The lubrication pump must be energized and must supply oil to the cutter blades before the cutting blades are engaged with metal. The schematic below is a simple design for such a system. Note how the second push-button system derives its power from a wire tapped into the first push-button system after the start button. In this system, for system 1 to provide power to system 2, system 1 must be energized. If system 1 is de-energized for some reason, system 2 will also automatically shut down.

Materials and Equipment

    (1)—12 volt dc power supply

    (2)—12 volt DPDT relay

    (2)—9–18 volt dc motor

    (2)—normally open (NO) push-button switch

    (2)—normally closed (NC) push-button switch

Procedure

    Step 1. Gather all materials required for this activity.

    Step 2. Connect the circuit according to the following schematic. Have your instructor check your work before energizing the circuit.

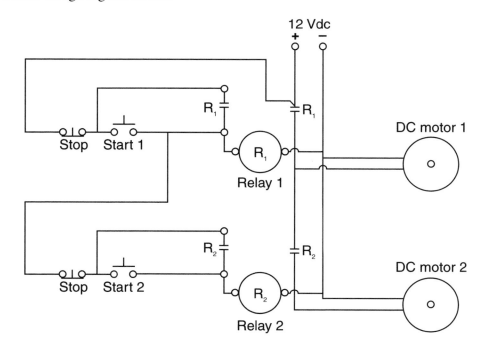

Step 3. Energize the circuit and check for proper sequencing as described in the introduction of this activity.

Step 4. Clear your work area. Properly store equipment and supplies.

Forward and Reverse Motor Control System

Name_____ **Score**_____

Date_____ **Class/Period/Instructor**_____

Introduction

A dc motor can be reversed by reversing the polarity applied to the armature circuit. In this lab activity, you will wire a forward and reverse control system for a dc motor. Two relays and two selector switches (SPDT) are utilized to accomplish this task. (Note: If only DPDT switches are available, they can be used instead. Simply use one half of each DPDT switch as your SPDT switches.)

This circuit must be arranged in such a way that both relays are not capable of being energized at the same time. If both relays were to energize at the same time, a deadly direct short circuit would occur. If properly wired, a system using two SPDT switches wired in series will prevent the accidental energizing of both relays at the same time. The first switch controls all power to both relays, while the second switch determines which relay is to be energized—forward or reverse. When building this circuit, carefully follow the schematic.

Materials and Equipment

    (1)—12 volt power supply

    (2)—12 volt DPDT relays

    (1)—9–18 volt dc motor

    (2)—SPDT (or DPDT) switches

    (1)—breadboard

Procedure

Step 1. Gather all materials required for this activity.

Step 2. Connect only the switches and relay as shown in the following schematic. Omit the motor supply circuit until the relay control circuit is tested. Verifying the relay control circuit before connecting the motor supply circuit will help to prevent an accidental short circuit on the power circuit.

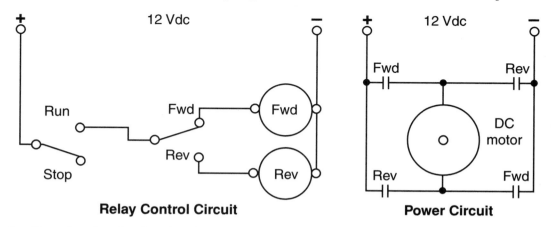

Relay Control Circuit   **Power Circuit**

Step 3. Energize the circuit only after your instructor approves it. Test the controls to see that the relay action occurs according to the switches. Make sure that both relays cannot be energized at the same time. Once you have verified that both relays will not close at the same time, proceed to Step 4.

(Continued)

Step 4. Connect the remaining part of the schematic that provides power to the motor. Have your instructor check it before going to Step 5.

Step 5. Energize the circuit and test the switching action to verify the circuitry. If the circuit fails to operate properly, check with your instructor.

Step 6. Clear your work area. Properly store equipment and supplies.

Transformers

Student Activity Sheet 12-1

Review

Name _____ **Score** _____

Date _____ **Class/Period/Instructor** _____

Complete the following sentences by filling in the missing word or words.

1. A transformer is a device that transfers energy from one circuit to another using _____ _____.

    1. _____

2. The input winding of a transformer is called the _____, while the output side of a transformer is called the _____.

    2. _____

    _____

3. _____ is the ability to produce electrical energy in a conductor without making physical contact with it.

    3. _____

4. The transferring of electrical energy from the primary to the secondary side of a transformer is an example of _____ induction. Creating energy in opposition to the input of the primary side of the transformer is called _____ induction.

    4. _____

    _____

5. Finish the transformer turns ratio formula below.

    $$\frac{E_P}{--} = \frac{I_s}{--} = \frac{}{--}$$
    $$\qquad\quad N_s$$

    5. _____

    _____

    _____

6. A transformer that has more turns of wire on the primary than on the secondary is called a(n) _____-_____ transformer.

    6. _____

7. A transformer that has more turns of wire on the secondary than on the primary is called a(n) _____-_____ transformer.

    7. _____

8. In a step-down transformer, the higher voltage is on the _____ side while the lower voltage is on the _____ side.

    8. _____

    _____

(Continued)

9. Transformer power is expressed as _____.

9. _____

10. In a step-down transformer, the higher current will be on the _____ side while the lower current will be on the _____ side.

10. _____

_____

11. Most transformer calculations are based on the assumption that transformers are 100% efficient. (In reality, this is not true.) What are the three types of transformer losses.

_____

_____

_____

12. Total opposition to ac current in the primary of a transformer is referred to as _____.

12. _____

13. The combination of winding losses and self induction is referred to as _____.

13. _____

14. A(n) _____ operates similar to an ohmmeter but utilizes much higher voltage for operation.

14. _____

15. A(n) _____ has only one continuous winding.

15. _____

16. The two main connection types for three-phase transformers are _____ and _____.

16. _____

_____

17. The _____ transformer connection is a series connection. The _____ transformer connection is a parallel connection.

17. _____

_____

For questions 18 through 20, supply the information requested using the given transformers.

18. Step-down transformer

    Primary current = _____

    Secondary voltage = _____

    Secondary current = _____

Turns ratio 20–1

120 Vac   2 Ω

19. Step-up transformer

    Primary current = _____

    Secondary voltage = _____

    Secondary current = _____

Turns ratio 1–5

12 Vac   3 Ω

20. Step-down transformer

    Primary voltage = _____

    Secondary current = _____

    Secondary load resistance = _____

Turns ratio 50–1

(A) 2 amps   18 Vac

Single-Phase AC Transformer

Name _____ **Score**_____

Date_____ **Class/Period/Instructor**_____

Introduction

In this activity, you will become familiar with connecting a center tap transformer and establishing a ground for the secondary. You will simulate how a single-phase power transformer establishes a neutral conductor. You will also learn why the center tap is chosen to be grounded. You will see the effect of impedance on the primary of the transformer and how current on the neutral is determined.

Materials and Equipment

    (1)—120-volt ac supply

    (1)—120-volt test pigtail

    (1)—voltmeter

    (1)—ammeter

    (1)—ohmmeter

    (4)—12-volt lamps and lamp holders

    (1)—120-volt primary, 12.6-volt 1.2-amp secondary center tap transformer

    (1)—fuse holder and 1/4 amp fuse

    (1)—dual-trace oscilloscope with two probes

Procedure

Step 1. Gather all materials required for this activity.

Step 2. Construct the circuit that follows. The primary has two wires and the secondary has three. Check with your instructor for the identification of the primary and secondary. **Do not energize this circuit until your instructor has checked your circuit. You are using 120 volts to energize this circuit. If it is improperly connected, dangerous high voltage could be present.**

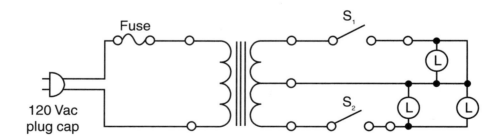

Step 3. After your instructor has approved the circuit, energize it by plugging the pigtail into a 120-volt ac source with S_1 and S_2 in the open position.

Step 4. Use the voltmeter to read the voltages on the secondary side of the transformer. Record the values in the spaces that follow. The secondary of the transformer should not yet be grounded.

       Input of S_1 to center tap   Voltage = _____ V

(Continued)

Input of S_1 to center tap   Voltage = _____ V

Input of S_1 to input of S_2 Voltage = _____ V

Input of S_1 to ground   Voltage = _____ V

Input of S_2 to ground   Voltage = _____ V

Input of center tap to ground   Voltage = _____ V

Question 1. Explain the reason for the differences in voltage readings on the secondary side of the transformer, especially to ground. _____

_____

_____

_____

_____

_____

Step 5. Connect the center tap to ground as indicated in the following schematic.

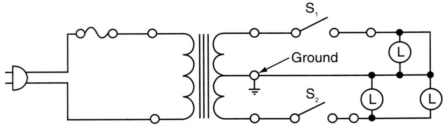

The center of the secondary is grounded.

Step 6. Take the voltage readings and record them below.

Input of S_1 to center tap   Voltage = _____ V

Input of S_2 to center tap   Voltage = _____ V

Input of S_1 to input of S_2 Voltage = _____ V

Input of S_1 to ground   Voltage = _____ V

Input of S_2 to ground   Voltage = _____ V

Input of center tap to ground   Voltage = _____ V

Step 7. Turn off the power to the transformer.

Question 2. Why are the voltage readings different when the center tap is grounded as compared to the voltage readings when the center tap is ungrounded? _____

_____

_____

_____

_____

_____

_____

(Continued)

Step 8. Calibrate and prepare the oscilloscope for dual-trace operation. Have the time sweep adjusted for 60 Hz and voltage amplitude for approximately 10 volts per major graticule. Set the horizontal reference for Channel 1 at two major divisions above center and Channel 2 at two major divisions below center.

Step 9. Connect the oscilloscope to the circuit as indicated below—Channel 1 to the input of S_1, and Channel 2 to the input of S_2. The ground connection is attached to the center tap of the transformer.

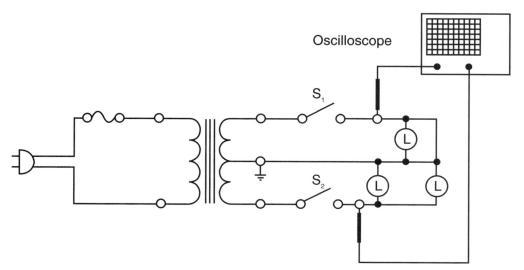

Step 10. Have your instructor check your oscilloscope connections. The oscilloscope is an expensive test device; be sure it is properly connected to avoid damage.

Step 11. Energize the transformer and adjust the oscilloscope until the two sine waves appear clearly and stationary on the display.

Step 12. Draw the waves on the grid below just as the two waves appear on the oscilloscope.

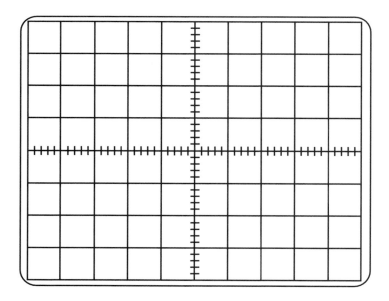

Question 3. What characteristics of the two sine waves appear to be similar and what characteristics appear to be different? (Amplitude, timing, polarity, etc.) _____

_____

_____

(Continued)

Question 4. What is the peak-to-peak value of each sine wave displayed?_____

_____

Peak-to-peak = _____ V

Step 13. Turn off the power. Disconnect and store the oscilloscope.

Step 14. Turn the power back on and close both S_1 and S_2.

Step 15. Observe and record the brightness of all lamps.

Step 16. Turn off the power and disconnect the center tap from the transformer to the lighting. This action will simulate an open neutral condition.

Step 17. Turn on the power and once again observe the brightness of the lamps.

Question 5. Describe the effect of an open center connection on the brightness of the lamps.

_____

_____

_____

_____

_____

_____

Step 18. Turn off the power. Now connect an additional lamp so that each line has an equal number of lamps. The load on each line will now be equal.

Step 19. Re-energize the circuit and observe the brightness of each lamp.

Question 6. Describe the effect of an open neutral with an evenly distributed load on each conductor.

_____

_____

_____

_____

_____

Step 20. Clear your work area. Properly store equipment and supplies.

Student Activity Sheet 12-3

Transformers—Determining Voltage, Current, and Turns Ratio

Name _____ **Score** _____

Date _____ **Class/Period/Instructor** _____

In this exercise, you will calculate the expected voltages, currents, or turns ratio based upon the information provided. Assume all the examples to have an efficiency rating of 100%.

1. What is the turns ratio of the transformer below? _____

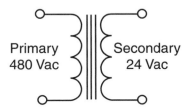

Primary
480 Vac

Secondary
24 Vac

2. The primary is connected to 120 volts ac and the transformer is constructed with a ten to one turns (10:1) ratio. What is the expected secondary voltage and current of the following transformer? What is the expected primary current?

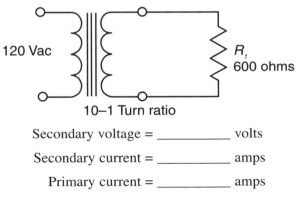

120 Vac

R_1
600 ohms

10–1 Turn ratio

Secondary voltage = _____ volts

Secondary current = _____ amps

Primary current = _____ amps

3. The following transformer has a 100 to 1 turns ratio. What are the expected primary voltage and current values?

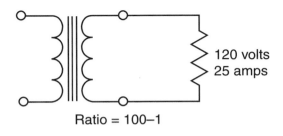

120 volts
25 amps

Ratio = 100–1

Primary voltage = _____ volts

Primary current = _____ amps

Student Activity Sheet 13-1

Review

Name_____ **Score**_____

Date_____ **Class/Period/Instructor** _____

Complete the following sentences by filling in the missing word or words.

1. The ac induction motor consists of two main parts. They are the _____, which revolves inside the _____ housing.

   1._____

   _____

2. The _____ consists of soft iron laminations and copper or aluminum bars running through, which connect on the ends.

   2._____

3. The three-phase induction motor commonly uses a(n) _____ _____ rotor.

   3._____

4. The speed of the magnetic field that rotates around the stator housing is called the _____ speed of the motor.

   4._____

5. The difference between the speed of the magnetic field rotation and the actual rpm of the rotor is called _____. It is usually expressed as a percentage.

   5._____

6. The three-phase _____ motor's rotor rotates at the same speed as the magnetic field of rotation. It has no slip.

   6._____

7. The three-phase _____ _____ motor has the ability to vary its speed under a load condition when a variable resistance is connected to the rotor circuit.

   7._____

8. The _____ _____ motor has its windings designed in pairs so that they can be connected either in series or parallel.

   8._____

9. The term _____-_____ means to make more than one phase out of a single-phase circuit.

   9._____

10. The _____ _____ induction motor uses a capacitor only in its start winding circuit.

    10._____

11. The capacitor start, capacitor run motor uses a(n) _____ in both the start winding and the run winding.

    11._____

12. In a capacitor start, capacitor run motor, a(n) _____ switch is used to disconnect the start winding from the power source.

    12._____

(Continued)

13. The ac motor known as the _____-_____ motor uses a commutator and brushes in addition to a standard rotor.

14. A motor that rotates only in small increments when the power is pulsed to it is called a(n) _____ motor.

15. Two types of motor protection are required in a dedicated line _____ _____ protection and _____ protection.

16. Branch circuit protection ranges from _____% to _____% of the motor's full load current.

17. The amount of overload that can be handled safely is called the _____ _____.

18. When a three-phase motor starts, the surge can be as high as _____% of full load current.

19. _____ protection is sized according to the motor's service factor.

20. A motor with a service factor of 1.1 can withstand an overload condition of _____%.

21. The most common cause of motor failure is _____ _____.

22. List three things that indicate a worn bearing.

_____

_____

_____

23. When a three-phase motor is running and one of the three fuses blows, you will have a(n) _____-_____ condition.

24. A(n) _____-_____ ammeter can be used to compare each of the three-phase conductors to see if they have equal current.

13. _____

14. _____

15. _____

_____

16. _____

_____

17. _____

18. _____

19. _____

20. _____

21. _____

23. _____

24. _____

Student Activity Sheet 13-2

Disassemble, Inspect, and Reassemble
a 1/4 Horsepower Motor

Name _____ **Score** _____

Date _____ **Class/Period/Instructor** _____

Introduction

In this activity, you will disassemble, inspect, and reassemble a 1/4 horsepower single-phase capacitor start motor. Many times the most practical way to verify motor winding damage is through visual inspection. The procedure of visual inspection should not result in damage to the motor housing or windings during disassembly or reassembly.

In this activity, you will learn the proper method to disassemble a motor, and after inspection, reassemble the motor safely without damage to the housing or windings. A typical motor has no front or back. Rather it has what is described using the terms *pulley end* and *opposite pulley end*. When disassembling a motor that you plan to be reassembling, realignment is critical. You will learn the proper technique in this lab to ensure reassembly of a motor without damage.

Materials and Equipment

(1) — 1/4 hp single-phase motor (Baldor or equivalent)

(1) — digital volt-ohmmeter (VOM)

(1) — set of socket drives

(1) — set of insulated screwdrivers

(1) — marker

(1) — plastic mallet

Procedure

Step 1. Gather all materials required for this activity.

Step 2. Quickly check the motor for bad bearings. Do this by checking the shaft for end play and lateral play by pulling on the motor shaft and moving it from side to side.

Step 3. Mark the motor housing at the seam of the pulley end and opposite pulley end. Put two marks across the pulley end and one mark across the other end. This will be a valuable guide when realigning the end bells to the stator housing during reassembly of the motor. If the end bells are not properly realigned with the stator, the through bolts may inadvertently be driven into the motor windings when they are inserted into the end bell. This will result in irreparable damage to the motor.

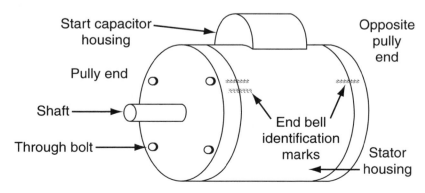

(Continued)

Step 4. Loosen all through bolts approximately one full turn before totally removing the nut from each bolt.

Step 5. Proceed to completely loosen all bolts one at a time.

Step 6. Remove the through bolts from motor housing.

Step 7. Remove the end bells from stator housing. If they are too snug, *gently* tap them with the plastic mallet. Excessive force can result in bending the stator housing.

Step 8. Gently remove the rotor from inside the stator housing.

Step 9. Visually inspect the stator winding. Look for dark or burnt spots on the windings. Also look for any signs of wear on the inside of the stator housing. Worn bearings will allow the rotor to touch the inside of the stator while the rotor rotates. A loose rotor spinning inside the stator will result in a shiny area on the inside of the stator caused by the rotor rubbing.

Step 10. Inspect the centrifugal switch mechanism mounted on the rotor shaft. Gently move the collar mechanism to ensure that it will slide along the shaft.

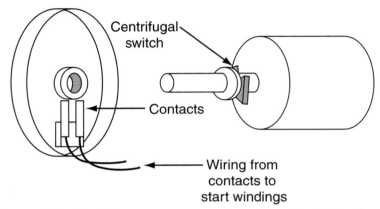

Step 11. Inspect the centrifugal switch contacts mounted on the end bell. Check to see if the contact points are excessively pitted or burnt.

Step 12. Remove the retaining screws from the capacitor housing to visually inspect the start capacitor. The start capacitor may appear to be bulging out at the top if it has experienced excessive heat. Another sign of excessive heat is discolored thick dark liquid seeping out of the vent hole. Also, look for discoloration such as burn marks where the motor leads connect to the capacitor.

Step 13. If the capacitor appears to be in good condition, place an insulated screwdriver shaft across the two connections. The capacitor is connected to the stator windings and centrifugal switch. This combination provides a shorted electrical path when in the normally closed (not running) position. If the centrifugal switch is good and the windings do not have an open, the capacitor will be completely drained of electrical charge. However, if there is a problem and the capacitor is not drained, there will be a substantial charge left in the capacitor. This charge could result in severe shock when the capacitor is handled. The shorting across the capacitor terminal with a screwdriver, or some other good conductor, will make sure that the capacitor is drained and safe to handle.

Step 14. Use a digital VOM for a quick check to determine if the capacitor is good. Set the digital VOM selector switch to mid-range ohms and connect it across the capacitor. You will see the numbers on the display flash from zero to infinity. If the reading on the meter holds steady at zero, the capacitor is shorted and defective.

Step 15. Reverse the leads of the VOM and closely watch the display. The numerical display should flash numbers across the screen until the meter once again reads infinity. You have just observed the charging and discharging action of the capacitor. The capacitor received a charge from the battery of the VOM.

(Continued)

Step 16. Now you are ready to reassemble the motor. Place the capacitor on the inside of the capacitor housing. Mount the housing to the stator housing.

Step 17. Place the rotor inside the stator housing.

Step 18. Slide each end bell over the shaft of the rotor. Be sure to align the marks you made across the seam of the end bells and stator housing. Do not force the end bells into alignment with the marks. Also, be careful not to pinch the wires from the centrifugal switch in the seam between the end bell and the stator housing.

Step 19. Carefully insert the through bolts into the end bell and through the stator housing. They should slide in and through the housing easily. Do *not* force them through the housing.

Step 20. Tighten the bolts and nuts using only your fingers until they are snug. Do *not* use the socket drives until all bolts and nuts are finger tight.

Step 21. Tighten each bolt only a few turns while at the same time rotating the shaft of the rotor. The rotor should turn freely while you are tightening the through bolts. If the rotor ceases to turn freely while tightening a bolt, stop immediately and tighten the bolt directly opposite on the end bell. The fit between the bearings in the end bells and the shaft is an extremely close tolerance. Overtightening any bolt can and will warp the end bell or damage the bearings. The rotor should turn freely at all times when retightening the through bolts. Tighten each bolt only a few turns before moving on to tighten another bolt. Keep moving around the end bell.

Step 22. Have your instructor check the motor before clearing your work area. When allowed to do so, clear your work area and properly store equipment and supplies.

Inductance and RL Circuits

Student Activity Sheet 14-1

Review

Name _____ **Score** _____

Date _____ **Class/Period/Instructor** _____

Complete the following sentences by filling in the missing word, words, or formula.

1. The property of an ac circuit that resists change in current is called _____.

    1. _____

2. A conductor shaped as a(n) _____ is a common form of inductor.

    2. _____

3. Some common sources of induction are _____ and _____.

    3. _____

    _____

4. The symbol for inductance is _____, and inductance is measured in a unit called the _____.

    4. _____

    _____

5. An example of _____ _____ is when two coils are in close proximity, and one coil transfers electrical energy to the other through a rising and collapsing magnetic field.

    5. _____

6. List three factors that affect the strength of self induction.

    _____

    _____

    _____

7. The _____ response is the response of the current and the voltage in a circuit after an instant change in applied voltage.

    7. _____

8. An example of _____ _____ is when a coil's magnetic field links with or cuts across the flux lines in another coil.

    8. _____

9. The letter symbol for reactance is _____. The letter symbol that denotes inductive reactance is _____.

    9. _____

    _____

10. Inductive reactance is measured in _____. The formula for inductive reactance is _____.

    10. _____

    _____

(Continued)

11. Inductive reactance increases when frequency _____ (increases/decreases) and decreases as frequency _____ (increases/decreases).

11. _____

_____

12. VAR stands for _____-_____-_____.

12. _____

13. Power factor is a comparison of _____ power and _____ power.

13. _____

_____

14. The formula for power factor is _____.

14. _____

15. The cosine of 20° is _____ and the cosine of 43° is _____. (Use the table in the appendix of the textbook.)

15. _____

_____

16. The total opposition to current in an ac circuit is called _____. Its letter symbol is _____.

16. _____

_____

17. The formula for impedance using the Pythagorean theorem method is _____.

17. _____

18. Finish the equation: True power = Apparent power × _____.

18. _____

19. In a parallel RL circuit, the current lags the voltage across the inductor by _____°.

19. _____

20. What is the formula for impedance in a parallel RL circuit?

20. _____

Inductance Characteristics

Name _____ **Score** _____

Date _____ **Class/Period/Instructor** _____

Introduction

In this activity, you will observe and record the typical characteristics of an inductive circuit using a typical audio transformer connected in series with a resistor. The audio transformer is rated for an impedance of 1 kΩ when connected to a frequency of 1 kHz. The audio transformer will act as an inductor in series with the resistor. You will observe the amplitude of the signal after the inductor and compare it to the original signal using a dual-trace oscilloscope. The phase shift produced by an inductor will become quite apparent in this activity.

Materials and Equipment

(1)—signal generator

(1)—audio output transformer

(1)—dual-trace oscilloscope

(2)—oscilloscope probes

(1)—breadboard

(1)—1 kΩ resistor, 1/4 W

Procedure

Step 1. Gather all materials required for this activity. Calibrate the oscilloscope; be sure to calibrate both channels.

Step 2. Construct the following circuit. You will connect only the two outside leads of the audio transformer in series with R_1. Do not use the center tap connection or the two leads of the output side of the transformer. All leads not used in this activity should be kept clear of the circuit. (You do not want the unused leads to short circuit. They must be kept in the open circuit condition.) Connect each lead not used to a different point in the breadboard. Do not energize the circuit until your instructor checks it.

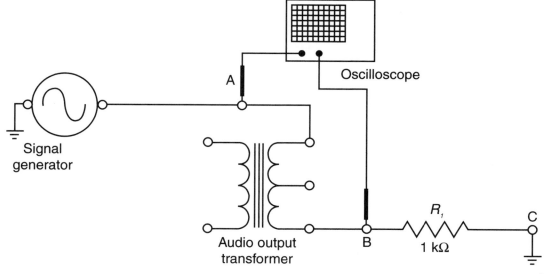

(Continued)

Step 3. Adjust the signal generator to produce a 1 kHz sine wave output. Record the amplitude of the signal for point A and point B. The connection of Channel A to point A will provide the monitoring of the applied frequency. The connection of Channel B to point B will provide a look at the input signal after it has been affected by the audio transformer. The induction caused by the transformer will affect the amplitude and cause a phase shift of the input signal. Draw the signal trace of both channels below as the frequency is increased from 1 kHz to 20 kHz. You will need to change the time division, or sweep rate, and the voltage division adjustments for both channels as you raise the frequency. Include the settings of the time sweep and voltage divisions for each reading to allow for easier interpretation of the trace drawings.

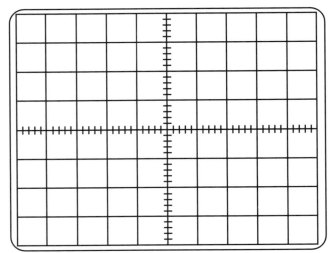

1 kHz input signal

Channel A and B = _____ Time/div.

    Channel A = _____ Peak-to-peak volts

    Channel B = _____ Peak-to-peak volts

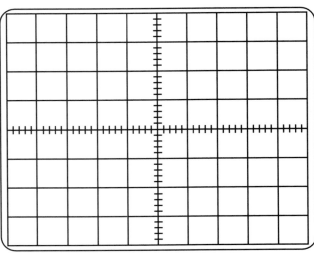

5 kHz input signal

Channel A and B = _____ Time/div.

    Channel A = _____ Peak-to-peak volts

    Channel B = _____ Peak-to-peak volts

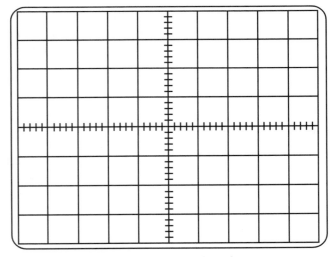

10 kHz input signal

Channel A and B = _____ Time/div.

    Channel A = _____ Peak-to-peak volts

    Channel B = _____ Peak-to-peak volts

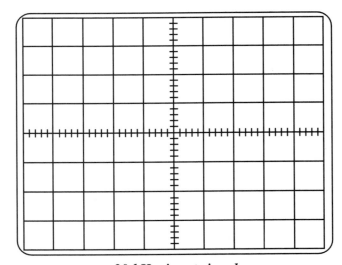

20 kHz input signal

Channel A and B = _____ Time/div.

    Channel A = _____ Peak-to-peak volts

    Channel B = _____ Peak-to-peak volts

(Continued)

Question 1. How was the amplitude of the signal trace of point A and B affected as the frequency increased?

_____

_____

_____

Question 2. Does the amplitude of the trace of Channel B inductor indicate an increase or decrease in circuit resistance? Explain. _____

_____

_____

_____

Student Activity Sheet 15-1

Review

Name _____ **Score** _____

Date _____ **Class/Period/Instructor** _____

Complete the following sentences by filling in the missing word or words.

1. A device that temporarily stores an electrical charge is a(n) _____.

   1. _____

2. A capacitor consists of two plates separated by a(n) _____.

   2. _____

3. The maximum voltage rating of a capacitor is called its _____ voltage.

   3. _____

4. The maximum voltage rating of the capacitor should exceed the _____ ac voltage being applied to the capacitor.

   4. _____

5. The unit of measure for capacitance is _____.

   5. _____

6. Capacitors are most commonly measured in two units. Name them.

   _____

   _____

7. The strength of a capacitor is determined by what three factors?

   _____

   _____

   _____

8. The strength of the capacitor _____ as the area of the plates decreases.

   8. _____

9. The capacitor strength decreases as the distance between the plates _____.

   9. _____

10. Most capacitors can be connected into a circuit without regard to polarity, except for _____ capacitors.

    10. _____

11. The time constant of a capacitor is the amount of time it takes for the capacitor to reach _____% of its full charge.

    11. _____

12. A capacitor will be fully charged at the end of _____ time constants.

    12. _____

13. Capacitance can be increased by connecting two or more capacitors in _____. Capacitance can be decreased by connecting two or more capacitors in _____.

    13. _____

    _____

(Continued)

14. Two 10 μF capacitors connected in parallel will equal _____ total.

14. _____

15. The letter symbol that represents capacitive reactance is _____.

15. _____

16. Capacitive reactance is measured in _____.

16. _____

17. The formula for capacitive reactance is _____.

17. _____

18. The capacitive reactance of a 20 μF capacitor connected to a 20 kHz source is equal to _____.

18. _____

19. Power factor is the relationship of _____ power to _____ power.

19. _____

_____

20. What is the power factor for a circuit that has a true power equal to 10 W and an apparent current equal to 2 A when connected to 12 V?

20. _____

Capacitor in Series with a Resistor

Name _____ **Score** _____

Date _____ **Class/Period/Instructor** _____

Introduction

In this lab activity, you will observe and record the typical characteristics of a series circuit consisting of a capacitor and a resistor. You will draw and interpret the trace of the circuit at different frequencies. This will allow you to observe the impedance of the circuit as the frequency changes.

Materials and Equipment

    (1)—signal generator

    (1)—5 pF capacitor, 50 V

    (1)—10 µF capacitor, 50 V

    (1)—1 kΩ resistor, 1/4 W

    (1)—dual-trace oscilloscope

    (2)—oscilloscope probes

    (1)—breadboard

Procedure

Step 1. Gather all materials required for this activity. Calibrate the oscilloscope.

Step 2. Construct the schematic below. Do not energize the circuit until your instructor has approved it.

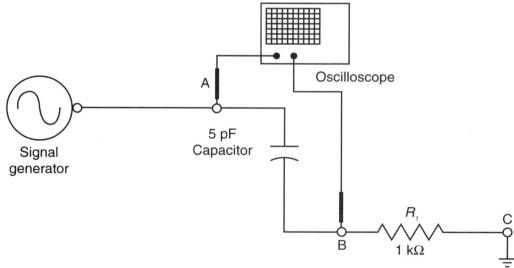

Step 3. Energize the circuit with the signal generator. Set the frequency of the input signal to match the frequencies indicated below each of the following oscilloscope screen drawings. The output voltage should be approximately 2 V peak-to-peak. Set the output voltage only once for this activity; do not readjust it for each frequency. Record the peak-to-peak voltage of both locations, 1 and 2, as seen on Channels 1 and 2. Draw the trace as it appears on the oscilloscope for each frequency indicated. You will need to readjust the time sweep setting for the oscilloscope each time the frequency of the signal generator is raised.

(Continued)

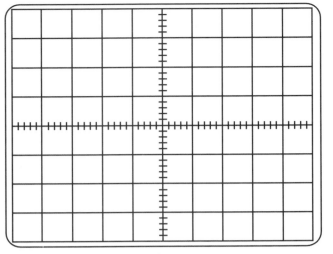

60 Hz

Channel 1 and 2 = _____ Time/div.

Channel 1 = _____ Peak-to-peak volts

Channel 2 = _____ Peak-to-peak volts

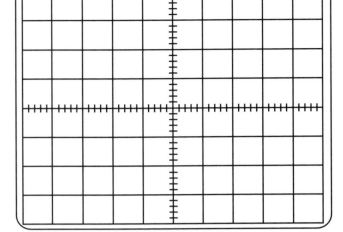

1 kHz

Channel 1 and 2 = _____ Time/div.

Channel 1 = _____ Peak-to-peak volts

Channel 2 = _____ Peak-to-peak volts

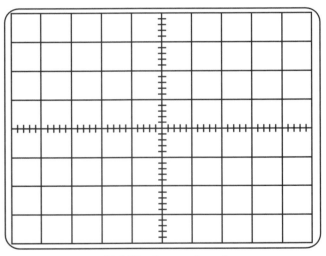

10 kHz input signal

Channel 1 and 2 = _____ Time/div.

Channel 1 = _____ Peak-to-peak volts

Channel 2 = _____ Peak-to-peak volts

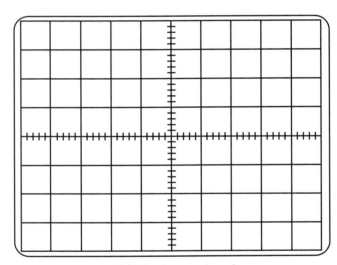

50 kHz input signal

Channel 1 and 2 = _____ Time/div.

Channel 1 = _____ Peak-to-peak volts

Channel 2 = _____ Peak-to-peak volts

Question 1. How did increasing the frequency affect the voltage drop across the capacitor and the resistor?

_____

_____

_____

_____

(Continued)

Advanced Electronic Circuits

Step 4. Insert the larger value capacitor in the circuit to replace the original capacitor. Repeat the experiment. Pay close attention to the sine wave pattern observed. You do not need to draw it.

Question 2. What effect does using a larger capacitor have on the circuit? _____

_____

_____

_____

Step 5. Clear your work area. Properly store equipment and supplies.

Capacitor Charge and Discharge Time Periods

Name _____ **Score** _____

Date _____ **Class/Period/Instructor** _____

Introduction

In this lab activity, you will verify the charge and discharge times of various capacitor and resistor combinations. The RC time constant is expressed in seconds, and it is derived from the amount of time required for a capacitor to become charged or discharged to 63.2% of the full potential applied to the capacitor. The time constant can be calculated by applying the formula $\tau = R \times C$, where τ represents time (in seconds), R represents resistance (in ohms), and C represents capacitance (in farads). The RC time constant is used in many electronic circuits, especially in timing circuits. You will apply the RC time constant several times in future activities.

Materials and Equipment

(1) — 10 Vdc power supply

(2) — SPST switches

(1) — 1 MΩ resistor, 1/4 W

(1) — 100 kΩ resistor, 1/4 W

(1) — 10 kΩ resistor, 1/4 W

(1) — 100 Ω resistor, 1/4 W

(1) — 6 μF capacitor, 50 V

(1) — 10 μF capacitor, 50 V

(1) — 47 μF capacitor, 50 V

(1) — voltmeter

(1) — stopwatch (or wristwatch with a second hand)

(1) — breadboard

Procedure

Step 1. Gather all materials required for this activity.

Step 2. Short all leads of each capacitor together to fully discharge each capacitor. Then, assemble the circuit as shown in the following schematic. Be sure to observe the proper polarity of each capacitor when constructing the circuit.

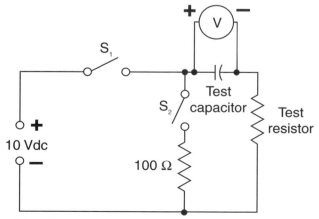

Step 3. Have your instructor check your setup before energizing the circuit.

Step 4. Examine the circuit you have set up. You must understand the purpose of the two switches before energizing the circuit. The two switches are designed to charge and discharge the capacitor. The 100 Ω resistor is installed in the circuit as a safety device in case you close S_1 while S_2 is closed. S_2 must be open while charging the capacitor. S_1 must be open when discharging the capacitor. The switches should not be closed at the same time.

Step 5. Energize the circuit and make observations by measuring the amount of time it takes to charge and discharge the capacitors with different values of resistance and capacitance. Test each of the combinations necessary to complete the chart. Fill in the values for the chart based upon your readings and calculations. The 100 Ω resistor used to discharge the capacitor will not affect the accuracy of the reading by more than 1%. Do not consider the 100 Ω resistor in your readings; use only the value inserted in the circuit as indicated by the chart.

$$\tau = R \times C$$
Time Period = Resistance × Capacitance
Full Charge = 5 Time Periods

Capacitor Value	Resistor Value	Actual Time of Full Charge	Calculated Time for Full Charge	Calculated Value for 1 Time Period
6 μF	100 kΩ			
6 μF	1 MΩ			
10 μF	100 kΩ			
10 μF	1 MΩ			
47 μF	100 kΩ			
47 μF	10 kΩ			

Question 1. Does the calculated time constant reflect the actual measurement of the time period in the activity?

_____

Question 2. How does increasing the resistance in the RC circuit affect the time period?

_____

Question 3. Which would you expect to have a shorter RC time period—a 10 Ω, 50 μF combination or a 100 Ω, 5 μF combination? Explain. _____

_____

_____

Step 6. Measure the discharge time of a fully charged capacitor. Pick a set of values from the chart and test the length of discharge time.

Question 4. How do the charge and discharge times of a capacitor compare? _____

_____

_____

Step 7. Clear your work area. Properly store equipment and supplies.

Tuned Circuits and RCL Networks

Student Activity Sheet 16-1

Review

Name_____ **Score**_____

Date_____ **Class/Period/Instructor**_____

Complete the following sentences by filling in the missing word or words.

1. An ac circuit connected to a resistor will produce _____ phase shift.

2. A resistor connected to an ac circuit will have _____ reactive power.

3. A resistor connected to an ac circuit will have its power consumption equal to _____ times _____.

4. If a coil had no dc resistance from the copper conductor, the current in the coil would _____ the voltage by _____ degrees.

5. A coil and a resistance connected together in an ac circuit will produce a current that _____ the voltage by an angle of _____ than 90 degrees.

6. A capacitor connected to an ac circuit will produce a current that _____ the voltage by approximately _____ degrees.

7. If a capacitor and resistor are connected to an ac circuit, the current will _____ the voltage by _____ than 90 degrees.

8. When the inductive reactance and the capacitive reactance are equal, the circuit is at a(n) _____ frequency.

9. The _____ through a series inductor-capacitor circuit is at a maximum at the resonant frequency.

10. When the frequency is at resonance, the voltage drops across the inductor and capacitor will be _____.

1. _____

2. _____

3. _____
   _____

4. _____
   _____

5. _____
   _____

6. _____
   _____

7. _____
   _____

8. _____

9. _____

10. _____

(Continued)

11. When an inductor is connected in parallel with a capacitor, and the resonant frequency is applied, the current will be at a(n) _____.

11. _____

12. Another name for a parallel capacitor and inductor circuit is the _____ circuit.

12. _____

13. The impedance of a parallel RCL circuit is at a(n) _____ at resonance while a series RCL circuit has a(n) _____ impedance at resonance.

13. _____

_____

14. The Q of the circuit is an indication of the _____ of the circuit.

14. _____

15. The bandwidth of a circuit is considered to be _____ percent above and below the resonance of the circuit.

15. _____

16. Filters are designed to either _____ or _____ frequencies.

16. _____

_____

17. A capacitor will _____ dc but _____ ac current.

17. _____

_____

18. A(n) _____-_____ filter will pass high frequencies while blocking low frequencies.

18. _____

19. A(n) _____-_____ filter will block high frequencies while passing low frequencies.

19. _____

20. A(n) _____ chart is handy for assisting in solving resonant frequency problems.

20. _____

Voltage in a Series RCL Circuit

Name _____ **Score** _____

Date _____ **Class/Period/Instructor** _____

Introduction
In this lab activity, you will vary the applied frequency of the power source and observe the effects on voltage drops across the components in a series RCL circuit. One of the most dramatic effects of RCL circuits is that the total sum of the voltage drops across the circuit components will exceed the value of the applied voltage. This is due to the phase shift caused by the inductor and the capacitor in the circuit. The pure resistance of the resistor causes no phase shift.

Materials and Equipment
> (1)—audio output transformer
>
> (1)—3.3 µF capacitor, 50 V
>
> (1)—1 kΩ resistor, 1/4 W (brown, black, red)
>
> (1)—digital voltmeter
>
> (1)—breadboard
>
> (1)—signal generator
>
> (1)—oscilloscope

Procedure
Step 1. Assemble all materials, and calibrate the oscilloscope.

Step 2. Connect the schematic that follows, but do not energize the circuit until your instructor has checked it for accuracy. Keep the unused terminals of the audio transformer in the clear. Do not connect them together or allow them to touch other conductors or ground.

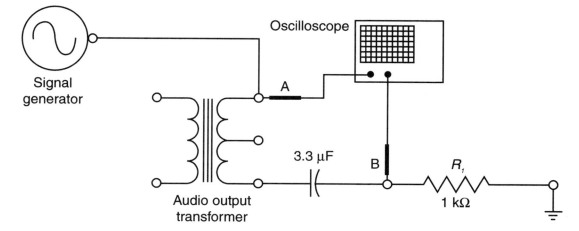

Step 3. Energize the circuit and adjust the signal generator until a 2 volt, 60 Hz sine wave is applied to the test circuit.

Step 4. Take readings across each component with the digital voltmeter and record the findings in the chart that follows.

(Continued)

Frequency of the applied 2 volts	Voltage drop at inductor	Voltage drop at capacitor	Voltage drop at 1 kΩ resistor	Total sum of voltage drops
30 Hz				
60 Hz				
120 Hz				

Step 5. Repeat Step 4 for an applied frequency of 30 Hz and 120 Hz. Adjust *only* the frequency and *not* the applied voltage of the circuit. Then record the voltage drops found.

Question 1. At which frequency did the inductor offer the greatest resistance?_____Hz

Question 2. At which frequency did the capacitor offer the greatest resistance? _____Hz

Question 3. At which frequency did the inductor offer the least resistance? _____Hz

Question 4. At which frequency did the capacitor offer the least resistance? _____Hz

Question 5. What would happen if the frequency of the source continued to rise? _____

_____

_____

_____

Question 6. What would happen if the frequency of the source was lowered to 1 Hz? _____

_____

_____

_____

Question 7. Based on the voltage drop across the 1 kΩ resistor, at which frequency was the current through the resistor the greatest? _____ Hz

Question 8. Which of the three experimental frequencies is closest to resonance, and why? _____

_____

_____

_____

_____

Step 6. Connect the dual trace oscilloscope across each of the components and observe the phase displacement when compared to the source.

Step 7. Clear your area and properly store all your materials.

Student Activity Sheet 16-3

Voltage in a Parallel RCL Circuit

Name _____ Score _____

Date _____ Class/Period/Instructor _____

Introduction

In this lab activity, you will experiment with a 1 kΩ resistor connected in series with a parallel LC circuit. An audio transformer will be connected in parallel with a 3.3 µF capacitor. You will observe and record the voltage drops across the circuit. After the voltages have been recorded as various frequencies are applied, you will make conclusions based on the apparent effect of the applied frequencies.

Materials and Equipment

    (1)—audio output transformer

    (1)—3.3 µF capacitor, 50 V

    (1)—1 kΩ resistor, 1/4 W (brown, black, red)

    (1)—signal generator

    (1)—digital voltmeter

    (1)—oscilloscope

Procedure

Step 1. Assemble all materials, and calibrate the oscilloscope.

Step 2. Connect the schematic that follows. Have your instructor check the circuit for accuracy before you energize.

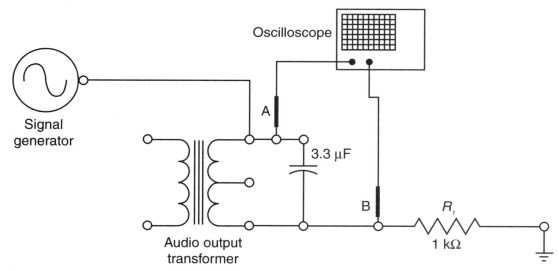

Step 3. Apply 2 volts at 60 Hz from the signal generator and verify the voltage using a digital voltmeter.

Step 4. Read the voltage drop across the resistor and then the inductor-capacitor parallel combination. Fill in the chart that follows. Repeat the voltage drop findings for 30 Hz and 120 Hz. Change only the frequency output of the signal generator, not the applied voltage.

(Continued)

Frequency of the applied 2 volts	Voltage drop across inductor-capacitor	Voltage drop across the 1 kΩ resistor	Combined total voltage drop
30 Hz			
60 Hz			
120 Hz			

Question 1. At what frequency did the resistor exhibit the largest current? Base your conclusion on the amount of voltage drop across the 1 kΩ resistor. _____ Hz

Question 2. At what frequency did the least current flow through the 1 kΩ resistor? Base your conclusion on the voltage drop across the 1 kΩ resistor. _____ Hz

Use the findings of this laboratory and the last laboratory activity (Student Activity Sheet 16-2, Voltage in a Series RCL Circuit) to complete the following questions.

Question 3. A _____ RCL circuit will have its greatest current at resonance while a _____ RCL circuit will have its least current value at resonance. (series/parallel)

Question 4. In general, a capacitor will block _____ frequencies and easily pass _____ frequencies, while an inductor will easily pass _____ frequencies and oppose _____ frequencies. (high/low)

Step 5. Following are three grid patterns of an oscilloscope. Draw the sine wave pattern displayed on a dual trace oscilloscope according to the frequency indicated. Identify the wave shape for the resistor and the inductor/capacitor circuit. Keep the resistor wave pattern centered in the scope display.

30 Hz

60 Hz

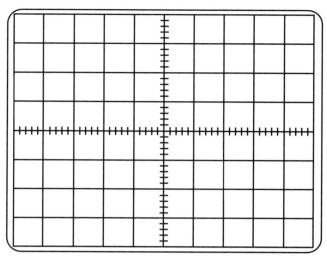

120 Hz

Question 5. Do these wave patterns verify the leading and lagging nature of the phase relationship of inductive and capacitive circuits?_____

Step 6.   Adjust the frequency output of the signal generator until both sine waves are in phase with each other.

Question 6. At what frequency is the circuit in resonance? _____Hz

Step 7.   Clean the work area and properly store all equipment and supplies.

Introduction to Semiconductors and Power Supplies

Student Activity Sheet 17-1

Review

Name_____ **Score**_____

Date_____ **Class/Period/Instructor**_____

Complete the following sentences by filling in the missing word or words.

1. The electrons in the outer most orbit ring of an atom are referred to as _____ electrons.

    1. _____

2. When the valence electrons bond to each other, the action is described as a(n) _____ bond.

    2. _____

3. A(n) _____ _____ structure is formed when valence electrons bond.

    3. _____

4. A(n) _____ is formed when an electron is removed.

    4. _____

5. In a semiconductor material, conduction of both _____ and _____ occurs.

    5. _____

    _____

6. A pure insulator material can be formed into a semiconductor material by a(n) _____ process.

    6. _____

7. A pure semiconductor is _____ while a doped semiconductor is _____.

    7. _____

    _____

8. A(n) _____ impurity has three electrons in the outer orbit while a(n) _____ impurity has five electrons in the outer orbit.

    8. _____

    _____

9. An atom that donates an extra electron in the doping process is classified as a(n) _____ impurity and will conduct electricity by electrons.

    9. _____

10. When a trivalent material is used in the doping process, a(n) _____ impurity is created, and conduction of electricity is by _____.

    10. _____

    _____

11. A material that conducts electricity is a(n) _____ type of crystal, while a material that conducts electricity by holes is considered a(n) _____ type of crystal.

    11. _____

    _____

12. A(n) _____ will conduct electrons in one direction, while blocking the flow in the opposite direction.

    12. _____

(Continued)

13. A diode consists of two types of semiconductor material, _____ and _____.

14. In a diode, two semiconductor materials are joined at the _____ barrier.

15. In a P-type material, electrical energy is conducted by _____ and in the N-type material by _____.

16. When a diode is forward biased, the positive side of the source is connected to the _____ crystal material, while the negative side is connected to the _____ material.

17. When a diode is reversed biased, the positive side of the source is connected to the _____ crystal material, while the negative side of the source is connected to the _____ material.

18. A(n) _____ diode may be used as a voltage regulator.

19. The abbreviation LED stands for _____ _____ _____.

20. In an LED, electrons and holes combine at the barrier to form a(n) _____, which is a particle of light.

21. The abbreviation PIV stands for _____ _____ _____.

22. A diode can be tested by using a(n) _____.

23. A transformer that has no electrical connection between the secondary and primary is called a(n) _____ transformer.

24. A connection to the midpoint of the secondary windings on a transformer is called a(n) _____ tap.

25. Changing ac current into dc current is called _____.

26. A(n) _____-_____ rectifier only converts one half of the sine wave to a dc current, which results in only half of the ac power being rectified.

27. A(n) _____-_____ rectifier converts both halves of the sine wave to a dc current.

28. Four diodes can be connected together to create a(n) _____ _____ circuit.

29. A rectification system that does not use an isolation transformer is called a(n) _____ _____ bridge circuit.

30. Connecting an oscilloscope directly to a line operated bridge rectifier will result in a(n) _____ _____.

31. To change the pulsating dc output of a rectifier to a smooth dc output, a(n) _____ network must be used.

32. _____ is the movement above and below the average dc output voltage and is expressed as a(n) _____.

33. A π filter network is composed of _____ and _____.

34. A load resistor is used as a(n) _____ and to improve _____.

13. _____

_____

14. _____

15. _____

_____

16. _____

_____

17. _____

_____

18. _____

19. _____

20. _____

21. _____

22. _____

23. _____

24. _____

25. _____

26. _____

27. _____

28. _____

29. _____

30. _____

31. _____

32. _____

_____

33. _____

_____

34. _____

_____

Diode Characteristics

Name_____ **Score**_____

Date_____ **Class/Period/Instructor**_____

Introduction

This lab activity will help you become familiar with a quick diode test and diode operation characteristics. A diode is a simple semiconductor that conducts current in only one direction. It consists of an N-type material joined to a P-type material. The color band on the diode usually denotes the cathode.

Materials and Equipment

    (1)—diode #1N4001

    (1)—diode #1N4005

    (1)—diode #1N5402

    (1)—diode #IN34A (germanium)

    (1)—1 kΩ resistor, 1/4 W (brown, black, red)

    (1)—SPST switch

    (1)—multimeter

    (1)—6 Vdc power supply

Procedure

Step 1. Assemble all materials.

A quick test to see if a diode is operating properly can be accomplished with an ohmmeter. Since the diode is made from two semiconductor materials joined together, resistance should be low, (less than 500 Ω) in one direction and relatively high (100 kΩ or more) in the opposite direction.

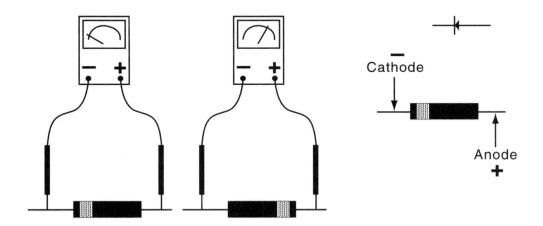

To test each diode simply connect the ohmmeter leads to the diode and then reverse the leads and read the meter again. Keep the meter range at or above the R × 100 range. Lower ranges such as R × 1 and R × 10 can allow too much current from the meter to be applied to the diode and destroy the diode.

Step 2. You will test each diode using an ohmmeter and record your findings in the chart below.

Diode type	Resistance value + to anode – to cathode	Resistance value – to anode + to cathode
1N4001		
1N4005		
1N5402		
Germanium #IN34A		

Question 1. How do the resistance values compare to each other? _____

_____

_____

A diode will conduct current in one direction and block it in the opposite direction. When the diode is conducting in a forward bias condition, there will be a slight voltage drop. Next you will measure and compare the voltage drop of different diodes. Pay particular attention to #IN34A, the germanium diode. All the other diodes being tested are silicon type diodes.

Step 3. Construct the circuit that follows to test each diode's voltage drop characteristics. Have your instructor check your circuit for accuracy before you begin testing.

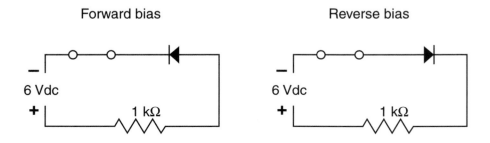

Step 4. Test each diode for forward and reverse bias voltage drop. Using a voltmeter, take a voltage drop reading across the diode and resistor and then record below in the chart. Next reverse the diode connections and repeat taking the voltage readings and then insert the reverse bias readings in the chart below.

Diode Type	Forward Bias Voltage Drop		Reverse Bias Voltage Drop	
	Diode	Resistor	Diode	Resistor
1N4001				
1N4005				
1N5402				
Germanium #IN34A				

Question 2. The average voltage drop for the first three diodes equals _____ volts, while the germanium diode voltage drop equals _____ volts.

Question 3. How did the diode voltage drops compare to one another when the diode was connected as reverse biased? _____

_____

_____

Question 4. What voltage reading would you expect for a diode that is open? _____

_____

_____

Question 5. What voltage reading would you expect for a normally functioning diode?_____

_____

_____

Step 5. Clear your work area. Properly store equipment and supplies.

Student Activity Sheet 17-3
Half-wave and Full-wave Power Supplies

Name_____ **Score**_____

Date_____ **Class/Period/Instructor**_____

Introduction

In this lab activity, you will build a power supply, both a half-wave supply and a full-wave supply, using a center tap transformer and diodes. You will also filter the supplies using capacitors. You will also need to review the concepts of *peak, average,* and *effective voltage.*

Materials and Equipment

(1)—center tap transformer, primary 120 V, secondary 6.3 V–6.3 V, 3 A

(4)—1N4005 diodes

(1)—10 μF capacitor, 35 Vdc

(1)—47 μF capacitor, 35 Vdc

(1)—1000 μF capacitor, 35 Vdc

(1)—2.2 kΩ resistor (red, red, red)

(2)—SPST switches

(1)—120 Vac power supply

(1)—oscilloscope

(1)—multimeter

Procedure

Step 1. Assemble all materials.

Step 2. Construct the following circuit and have your instructor check it for accuracy before you energize this circuit. **Caution!** *You will be using a 120 Vac power source. Pay close attention to this lab, and do not touch any exposed connections with your bare hands. Keep your fingers away from the tip of the meter test leads to avoid shock.*

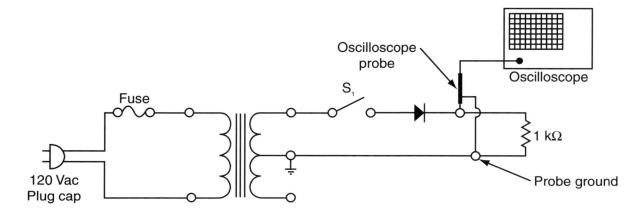

Step 3. Energize the circuit. Draw the waveforms before and after rectification on the screens that follow. Set your oscilloscope controls to approximately 1 mS time/div. and 2 volts.

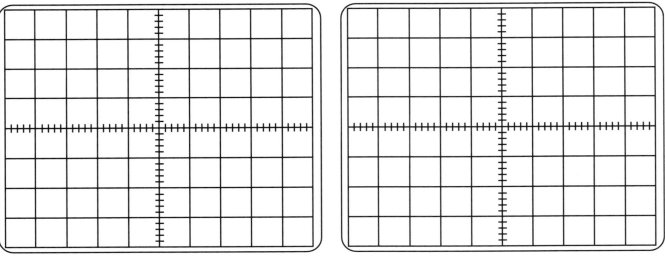

ac wave pattern   half-wave dc rectified pattern

Question 1. What is the peak-to-peak voltage of the ac wave? _____

Question 2. What is the peak-to-peak voltage of the dc wave?_____

Step 4. Turn off the power to the transformer.

Step 5. Connect the full-wave rectifier circuit that follows. Do not connect a capacitor to the circuit at this time. The capacitor will be installed after initial tests have been run to serve as a reference. Have your instructor check your circuit before energizing.

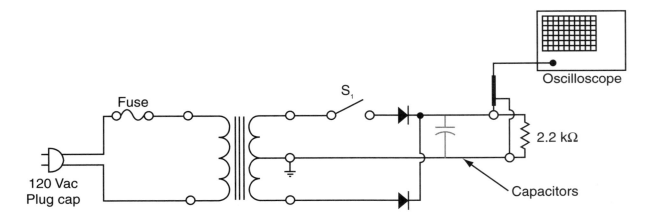

Step 6. Record the sine wave pattern of the oscilloscope for a the full wave rectifier at the load.

Peak-to-peak voltage = _____

Frequency = _____

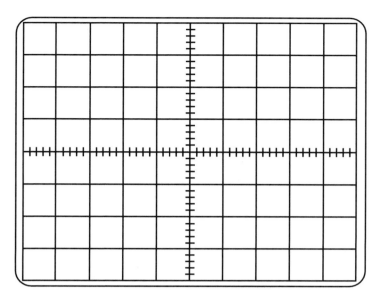

Step 7. Now you will start to filter the circuit. Filtering reduces output ripple of the power supply. With the oscilloscope connected across the load, add each capacitor one by one and observe the change in the pulsating dc wave pattern. Record the patterns on the screens that follow.

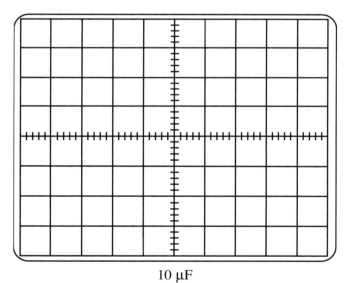

10 µF

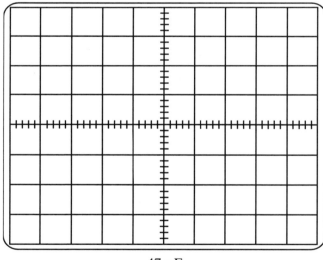

47 µF

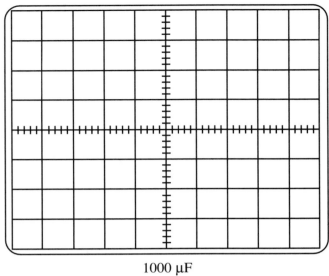

1000 μF

Fill in the requested values. Use a multimeter for the rms values and a oscilloscope for the peak values.

10 μF = _____ dc rms volts   10 μF = _____ oscilloscope peak volts

47 μF = _____ dc rms volts   47 μF = _____ oscilloscope peak volts

1000 μF = _____ dc rms volts   1000 μF= _____ oscilloscope peak volts

Question 3. If you suspected a power supply was not working properly, how would you troubleshoot it? What instruments would you use and why? _____

_____

_____

_____

_____

_____

_____

_____

Step 8. Clear your work area. Properly store equipment and supplies.

Student Activity Sheet 17-4

Full-Wave Bridge Power Supply

Name _____ **Score** _____

Date _____ **Class/Period/Instructor** _____

Introduction

In this lab, you will build a full-wave bridge rectifier using four diodes. This power supply can be built using an ac source without a transformer. In this lab, we will use a transformer to provide isolation so that you can connect an oscilloscope without the risk of damage from connecting the scope ground to a hot ac conductor.

Materials and Equipment

(1)—center tap transformer, primary 120 V, secondary 6.3 V–6.3 V, 3 A

(1)—oscilloscope

(1)—multimeter

(1)—2.2 kΩ resistor, 1/4 W (red, red, red)

(4)—1N4005 diodes

(1)—DPDT switch

(1)—breadboard

Procedure

Step 1. Assemble all materials.

Step 2. Connect the circuit below. Have your instructor check your circuit before energizing.

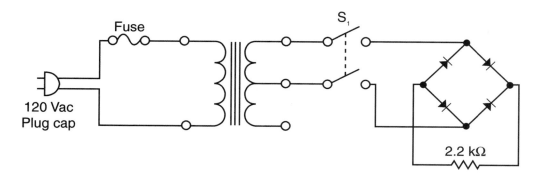

Step 3. Connect the oscilloscope across the load resistor and observe the sine wave pattern.

Step 4. Connect a multimeter across the load resistor and note the dc rms voltage.

Question 1. How does the dc rms voltage of the full-wave bridge rectifier compare to the dc rms voltage of the center tap transformer two diode full-wave bridge circuit in Student Activity Sheet 17-3?

_____

_____

_____

_____

Question 2. How is polarity determined at the output of the power supply? _____

_____

_____

_____

Question 3. What would happen if all four diodes were reversed in their positions? _____

_____

_____

_____

Question 4. What would happen if only one diode was reversed? _____

_____

_____

_____

Step 5. Clear your work area. Properly store equipment and supplies.

Zener Diode Characteristics

Name _____ **Score** _____

Date _____ **Class/Period/Instructor** _____

Introduction

The best explanation of how a zener diode operates is to observe the diode in a circuit with a varying voltage applied to the diode circuit. Each zener diode has a particular conducting voltage and will not conduct until that particular voltage is reached. The zener for this lab is rated at 6.2 volts, which means it should not conduct until 6.2 volts is reached. Zener diodes are used to regulate voltage levels in equipment or power supplies.

Materials and Equipment

(1)—6.2 V zener diode 1N4735

(1)—47 kΩ resistor, 1/4 W (yellow, violet, orange)

(1)—2.2 kΩ resistor, 1/4 W (red, red, red)

(1)—0-12 V variable dc supply

(1)—multimeter

(1)—breadboard

Procedure

Step 1. Assemble all materials.

Step 2. Construct the circuit that follows, but leave out the zener diode. You need to perform a set of tests before adding the diode. Have your circuit checked by your instructor before energizing.

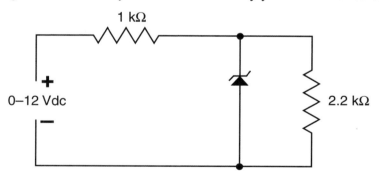

Step 3. Set the power supply voltage for 4 Vdc. Measure the voltage across the output, the 2.2 kΩ resistor. Adjust the power supply to each of the voltages listed in the chart that follows. Measure the voltage across the 2.2 kΩ resistor for each new input voltage. Record all of your readings in the chart.

Before Zener Diode in Circuit

Source voltage (V)	4	5	6	7	8	9	10	11	12
2.2 kΩ voltage (V)									

Step 4. Install the zener diode across the 2.2 kΩ resistor as shown in the schematic. Repeat step 3 and record your findings in the chart that follows. The zener diode is installed as a reverse bias diode. In forward bias it will act like a normal diode. Remember, using a zener diode as a voltage regulator takes advantage of the reverse bias breakdown at the barrier junction. Have your instructor check your circuit before energizing.

After Zener Diode in Circuit

Source voltage (V)	4	5	6	7	8	9	10	11	12
2.2 kΩ voltage (V)									

Question 1. At what source output voltage did the zener diode start to maintain the same voltage across the

2.2 kΩ resistor? _____ volts dc.

Question 2. Can the zener diode be used to maintain a constant voltage applied to a circuit?_____

_____

Question 3. How would you connect a zener diode across a power supply output to maintain the output

voltage under changing load conditions? Draw the circuit below.

Step 5. Clear your work area. Properly store equipment and supplies.

Voltage Doubler Power Supply

Name _____ **Score** _____

Date _____ **Class/Period/Instructor** _____

Introduction

A voltage doubler power supply can increase the output level of a power supply. Carefully study the schematic diagram in Step 2. Take note of the way the two capacitors are connected into the circuit. Each capacitor will store a charge from each diode. D_1 will conduct while D_2 blocks current. D_1 will charge C_1. On the second half of the ac cycle, will conduct while D_1 blocks current. D_2 will charge C_2. Each capacitor stores one half of the total voltage. The combined effect of the two charged capacitors is applied across the load resistor.

Materials and Equipment

(1)—center tap transformer

(2)—diodes 1N4001

(2)—10 µF capacitors, 35 V

(1)—2.2 kΩ resistor, 1/4 W (red, red, red)

(1)—multimeter

(1)—breadboard

Procedure

Step 1 Assemble the materials needed.

Step 2. Connect the circuit shown in the schematic that follows. Have your instructor check your work before you energize the circuit.

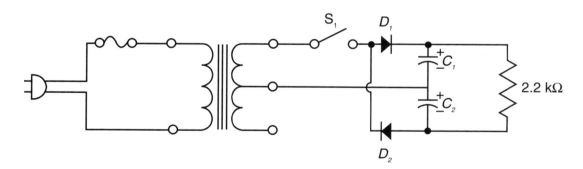

Step 3. Take a voltage reading across the resistor and record it. Resistor voltage = _____

Step 4. Take a voltage drop reading across C_1 and C_2. Record them. C_1 = _____ volts;

C_2 = _____ volts

Question 1. Does the voltage drop across the resistor equal twice the voltage across either capacitor?

_____

Step 5. Clear your work area. Properly store equipment and supplies.

Student Activity Sheet 17-7

Positive and Negative Power Supply

Name _____ **Score** _____

Date _____ **Class/Period/Instructor** _____

Introduction

At times, it is necessary to provide a positive and a negative power source for a circuit. Some electronic components require a positive and an equal negative supply for certain applications. One of these components is the operational amplifier, which will be covered in an upcoming chapter. The circuit below is similar to the voltage doubler circuit. It has been modified to establish a ground, or neutral. The neutral is, as the name implies, neither positive nor negative. The ground helps to establish it as a reference point. After completing the circuit, you will proceed to take voltage readings to find out if an incandescent lamp will operate on negative power as well as positive power. Can you guess?

Materials and Equipment

    (2)—diodes 1N4001

    (1)—center tap transformer

    (2)—10 µF capacitors, 35 V

    (2)—2.2 kΩ resistors, 1/4 W (red, red, red)

    (2)—12-volt lamps

    (1)—multimeter

    (1)—breadboard

Procedure

Step 1. Assemble the materials needed.

Step 2. Connect the circuit in the schematic that follows. Be sure to have your instructor check your work before you energize the circuit. Also, be sure to connect the capacitors with the correct polarity.

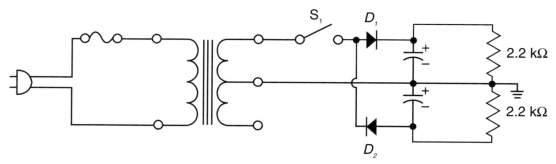

Step 3. Take a voltage reading across each resistor. Record your readings.

    Top resistor = _____ V

    Bottom resistor = _____ V

Step 4. Take a voltage drop reading across both resistors. Record your reading.

    Both resistors = _____ V

Question 1. Describe the polarity across the top and bottom resistors.

Step 5. Connect a lamp in parallel with each resistor and see if they both will light.

Question 2. Did the two lamps shine at the same brightness?_____

Step 6. Clear your work area. Properly store equipment and supplies.

Light Emitting Diodes—LEDs

Name _____ **Score** _____

Date _____ **Class/Period/Instructor** _____

Introduction

In this lab activity, you will experiment with the light emitting diode (LED). LEDs come in a variety of colors such as red, yellow, green, and blue. Some are in the infrared region and are not visible to the human eye. These appear clear.

When an LED is connected in forward bias, it will light up. The LED is used as an indicator light in many electronic circuits. An LED must always be connected with a current limiting resistor. Some forms of LEDs have the resistor built into them to limit current. Unless you know for certain that an LED contains its own resistor, always assume that an LED requires one.

The typical LED requires a resistor in the range of 300 to 1000 ohms. LEDs are usually rated in current value. You must calculate the size of resistor required to limit the current. The LED below (part #276-041A) has a maximum forward current value of 20 mA. For this lab, it will be necessary for you to calculate the lowest value of resistor allowed for this LED.

Materials and Equipment

(1)—12 Vdc and ac power supply

(2)—resistors, size to be determined

(1)—330 Ω resistor, 1/4 W (orange, orange, brown)

(2)—red LEDs

(1)—LED seven segment display

(1)—breadboard

Cathode ——— −

Anode ——— +

LED

Procedure

Step 1. Calculate the resistor size needed for the 20 mA LED when connected to 6 volts and 12 volts. The forward voltage across the LED is 2 volts. Use the next larger standard size resistor for your answer.

Resistor size when using 6 volts = _____ ohms

Color bands for this resistor = _____ _____ _____

Resistor size when using 12 volts = _____ ohms

Color bands for this resistor = _____ _____ _____

Step 2. Have your instructor check your calculations before proceeding further.

(Continued)

Step 3. Connect the circuit below. Remember that the cathode is the short lead and is connected to the negative side of the supply.

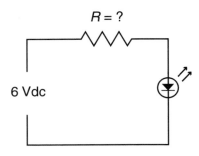

Question 1. Does the LED light? _____

Step 4. Reverse the leads of the LED so that it is connected in reverse bias, and observe.

Question 2. Does the LED light when reverse biased? _____

Step 5. Connect the circuit to the 6 Vac power supply, and energize. Observe the LED.

Question 3. Does the LED light? _____

Step 6. Reverse the leads on the LED to see if it will light again.

Question 4. Why does the LED light when provided an ac voltage with leads in either direction, but on dc the LED must be forward biased? Explain. _____

_____

_____

Step 7. Measure the voltage drop across the LED and enter that value here.

LED volt drop = _____

Question 5. Will the voltage drop be the same or increase when a higher voltage is applied? Explain your answer.

_____

_____

_____

Step 8. Repeat steps 3 through 7 for using 12 V. Remember to change the resistor value for the 12 volt calculated value. Verify your answer for Question 5.

In the second part of this lab activity, you will connect a seven segment LED display. It is a commonly used visual indicator for such things as voltage, current, resistance, speed, weight, and distance. In this part of the laboratory, you will become familiar with how the segments are connected and identified. A direct application of the LED display will come in a later lab activity involving a digital counter. Look at the pin diagram of the seven segment LED display. Each pin is connected to one of the seven segments that comprise the display unit. Each segment is an individual LED arranged in a figure eight pattern. Displays also come in matrixes and bar graph styles.

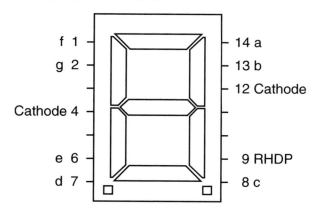

Step 9. Set the source voltage to 6 volts dc.

Step 10. Connect either cathode to the negative side of the supply.

Step 11. Connect the 6-volt positive source through the 330-ohm resistor to each of the pins marked a, b, c, d, e, f, g. Observe which segment lights and shade that portion in the illustration below. There is one illustration for each of the seven segments.

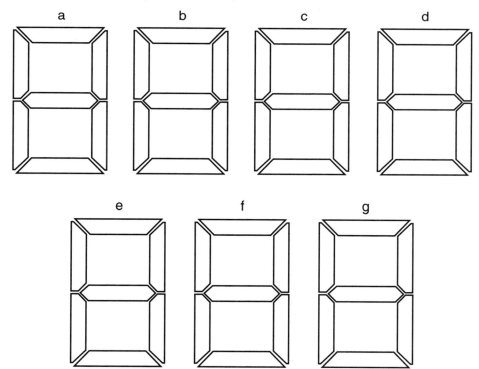

Question 6. Could more than one segment be lit at the same time? Explain._____

_____

_____

Step 12. Clear your work area. Properly store equipment and supplies.

Tubes, Transistors, and Amplifiers

Student Activity Sheet 18-1

Review

Name_____ **Score**_____

Date_____ **Class/Period/Instructor**_____

Complete the statements below by filling in the missing word or words.

1. The vacuum tube has been replaced by _____ circuits.

    1. _____

2. _____ _____ takes place when metals are heated to a sufficient temperature to release electrons.

    2. _____

3. The emitter in a vacuum tube is called the _____.

    3. _____

4. The vacuum tube diode construction consists of a(n) _____ and a(n) _____.

    4. _____
    _____

5. The _____ is heated and releases electrons that are collected on the positive _____.

    5. _____
    _____

6. When the action of increasing the applied voltage will not cause more electrons to be collected on the anode, the tube has reached its _____ _____.

    6. _____

7. A three element tube is called a(n) _____.

    7. _____

8. The electron flow between the cathode and anode is controlled by the _____ _____.

    8. _____

9. The _____ _____ of the tube is reached when a negative voltage is applied to the control grid and electron flow ceases.

    9. _____

10. A four element tube is called a(n) _____.

    10. _____

11. A four element tube contains a cathode, control grid, plate, and a(n) _____ _____.

    11. _____

12. A five part tube is called a(n) _____.

    12. _____

13. In a five part tube, secondary emission is prevented by the _____ grid.

    13. _____

14. The abbreviation CRT stands for _____ _____ _____.

    14. _____

15. The position of the CRT electron beam is controlled by _____ _____ coils.

    15. _____

(Continued)

16. The fluorescent coating on the screen of the CRT glows when struck by _____.

16. _____

17. Transistors are able to _____ current, create desired ac _____, and can be used as _____ devices.

17. _____

_____

_____

18. The two types of bipolar transistors are _____ and _____.

18. _____

_____

19. The three main parts of the transistor are the _____, the _____, and the _____.

19. _____

_____

_____

20. The arrow in the schematic symbol of a transistor points toward the _____-type material.

20. _____

21. The _____ controls the flow of electrical energy through the emitter collector circuit.

21. _____

22. For a NPN transistor, the emitter is connected to the _____ polarity while the collector is connected to the _____ polarity.

22. _____

_____

23. The abbreviation FET stands for _____ _____ _____.

23. _____

24. The major electrical operation difference of a FET compared to a bipolar transistor is the FET is classified a(n) _____ device while the transistor is a(n) _____ device.

24. _____

_____

25. The abbreviation MOSFET stands for _____ _____ _____ _____-_____ _____.

25. _____

_____

26. The three main connection points on a FET are the _____, the _____, and the _____.

26. _____

_____

_____

27. A(n) _____ is an electrical device or circuit that controls a large electrical signal by applying a small electrical signal to the circuit or device.

27. _____

28. The ratio between input power and output power is called _____.

28. _____

29. Connecting the polarity for current flow from the emitter to the base is called _____ _____.

29. _____

30. The locations of barrier junctions in a transistor are between the _____ and the _____, and between the _____ and the _____.

30. _____

_____

_____

_____

(Continued)

31. For a transistor to operate properly, there must be both _____ and _____
biasing.

31. _____

_____

32. The three circuit configurations for transistor circuits are common _____,
common _____, and common _____.

32. _____

_____

_____

33. The three main classifications of amplifiers are class _____, class _____,
and class _____.

33. _____

_____

_____

34. Thermal runaway will cause _____ of the transistor.

34. _____

35. Coupling transistors together is called _____.

35. _____

36. Transistors may be connected together to increase amplification by _____
coupling, _____ coupling, _____ coupling, and _____ coupling.

36. _____

_____

_____

_____

37. The SCR is designed to handle large _____ currents while the triac is
designed to handle large _____ currents.

37. _____

_____

38. The _____ consists of P, N, P, N, layers of semiconductor.

38. _____

39. The three main terminals of an SCR are the _____, _____, and the _____.

39. _____

_____

_____

40. The three main leads on a triac are _____, _____, and _____.

40. _____

_____

_____

Basic Review of Transistors

Name_____ **Score**_____

Date_____ **Class/Period/Instructor**_____

This activity is designed to familiarize you with the essentials of transistors before attempting your first transistor laboratory activity.

Write the name of the part that corresponds with the numbered blank on the right.

1. _____

2. _____

3. _____

4. _____

5. _____

6. _____

Transistor A   Transistor B

Identify which transistor above is an NPN and a PNP type transistor.

7. Transistor A is _____.

8. Transistor B is _____.

7. _____

8. _____

Complete the statements below about the two transistors, A and B above, when connected as a common emitter amplifier.

9. On transistor A, connection point 3 should be connected to a _____ polarity while connection point 1 should be connected to a _____ polarity.

9. _____
_____

10. On transistor B, connection point 6 should be _____ polarity while connection 4 should be _____ polarity.

10. _____
_____

11. In reference to transistor A, the greatest current will be from point _____ to point _____.

11. _____
_____

12. In reference to transistor B, the greatest current will be from point _____ to point _____.

12. _____
_____

13. In reference to transistor A, an ac signal applied to point _____ will control the current from point _____ to point _____.

13. _____
_____
_____

14. In reference to transistor B, an ac signal applied to point _____ will control the current from point _____ to point _____.

14. _____
_____
_____

Identifying and Testing Transistors

Name_____ Score_____

Date_____ Class/Period/Instructor_____

Introduction

At times, it is necessary to identify transistor leads and also to check the condition of a transistor. In this laboratory you will learn to identify a transistor as a PNP or an NPN, and you will learn to make a quick check with an ohmmeter to see if the transistor is shorted, open, or good. The first part of the lab explains the testing procedure you will follow. Then, you will test the MPS2222A transistor as an example of expected readings. You will then test 10 transistors to determine their type (NPN or PNP) and their condition.

Materials and Equipment

(1)—multimeter

(1)—breadboard

(1)—NPN transistor, MPS2222A

(10)—assorted transistors provided by instructor

Procedure

Step 1. Assemble all materials.

Step 2. Carefully read all of the following material before proceeding to Step 3. First, we will review the crystal structure of a transistor and the expected resistance values when read with an ohmmeter. Look at the drawing below. As you can see the NPN transistor consists of three layers of crystal as the name implies, N-type, P-type and again N-type. Look at the diagram of the pin arrangement of the MPS2222A. Take special notice of the way the diagram is labeled (viewed from bottom). It is easy to become confused when working between transistor diagrams and the actual project.

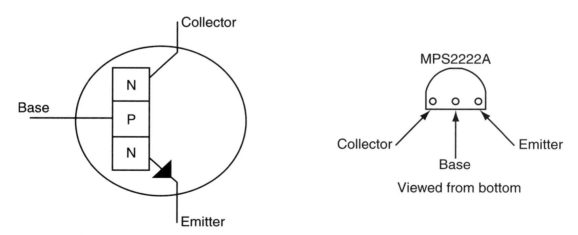

The base is common to both the emitter and collector. The base-to-emitter and base-to-collector junctions will have low resistance readings when the positive (red) lead is connected to the base, and the negative (black) lead is connected to the collector or emitter.

Low Resistance

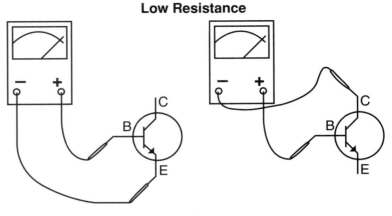

Take note of the relationship of the meter lead polarity and the transistor's pins. The NPN transistor is forward-biased, which causes the low resistance reading. A low resistance reading is considered approximately 500 to 1000 ohms.

High Resistance

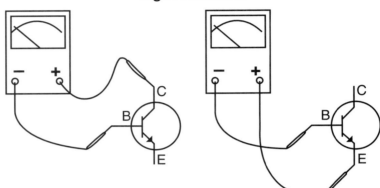

When the leads of the meter are reversed, the resistance reading will be quite high. A high resistance reading is well over one million ohms. In this position, the ohmmeter is connected in reverse-bias condition.

High Resistance

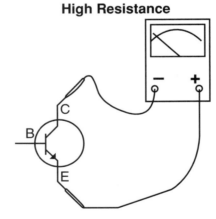

The collector to emitter readings will be quite high for either polarity reading. The transistor has two barrier regions. One barrier region is between the base and emitter and the other barrier region is between the base and collector. Reversing the applied polarity from the ohmmeter will not change the value of the ohmmeter reading. If the reading is low in either direction, then the transistor is defective.

Step 3. Set the range of your multimeter to at least the R × 100 ohms setting before taking any readings. A range selection of R × 1 or R × 10 may damage the transistor while it is being tested.

Step 4. Identify the three leads on the MPS2222A transistor. Be sure of the emitter, collector, and base lead locations. Remember to view the transistor from the bottom.

Step 5. Connect the ohmmeter to the MPS2222A transistor. Keep the positive lead connected to the base while moving the negative lead to the collector and emitter. Record the resistance values.

Resistance from base to collector (forward biased) = _____ ohms

Resistance from base to emitter (forward biased) = _____ ohms

Step 6. Now reverse the leads of the ohmmeter and repeat the readings taken in Step 5. The negative lead should be connected to the base and the positive lead moved between the collector and emitter. Record the resistance values.

Resistance from base to collector (reverse biased) = _____ ohms

Resistance from base to emitter (reverse biased) = _____ ohms

Step 7. Your readings should have been quite high for the readings in Step 6 and relatively low in Step 5. If this is not the case, check with your instructor.

Step 8. Connect the meter to the collector and emitter. First connect the positive lead to the collector and the negative lead to the emitter, and then connect the meter again but this time with the opposite polarity. Record the resistance values for both polarities.

First reading, collector to emitter = _____ ohms

Second reading, collector to emitter = _____ ohms

Question 1. In forward bias condition, would the resistance from the collector to the emitter be high or low?

_____

Question 2. When testing an NPN transistor, which ohmmeter lead would you connect to the base for a

forward bias condition? _____

Question 3. When testing a PNP transistor, which ohmmeter lead would you connect to the base? _____

Question 4. Why is the resistance value high between the emitter and collector? _____

_____

Step 9. Using an ohmmeter, you will now test the 10 remaining transistors to determine if they are NPN or PNP and if their condition is considered good or bad. Fill in the chart below as you test each transistor.

Transistor number	Type (NPN, PNP)	Condition (Good, Bad)

Question 5. Describe how the resistance readings of one good transistor relates to any other good transistor.

_____

_____

Step 10. Clear your work area. Properly store equipment and supplies.

Common Emitter Bias

Name_____ **Score**_____

Date_____ **Class/Period/Instructor**_____

Introduction

In this lab activity, you will establish a bias voltage for a transistor. The proper bias voltage is critical for the expected amplification to take place. In this lab, you will manipulate voltage at the base of the transistor using a 50 kΩ potentiometer.

Materials and Equipment

  (1)—0-12 variable power supply

  (1)—signal generator

  (1)—multimeter

  (1)—100 kΩ potentiometer

  (1)—50 kΩ potentiometer

  (1)—22 kΩ resistor 1/4 W (red, red, orange)

  (1)—4.7 kΩ resistor 1/4 W (yellow, violet, red)

  (1)—1 kΩ resistor 1/4 W (brown, black, red)

  (1)—MPS2222A NPN transistor (or equivalent)

  (1)—dual trace oscilloscope

  (1)—4.7 µF capacitor, 50 V

  (1)—LED (red)

  (1)—breadboard

Procedure

Step 1. Assemble all materials, and calibrate the oscilloscope.

Step 2. Connect the circuit below. Pay particular attention to the location of the emitter, collector, and base on the transistor. Do not energize the circuit until your instructor checks your work.

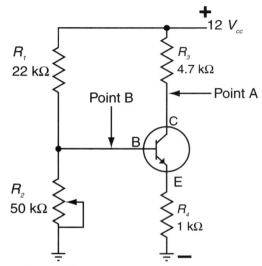

Step 3. Adjust the source voltage to 12 Vdc.

Step 4. Connect the oscilloscope to test point A, and observe the action on the screen while slowly adjusting the 50 kΩ potentiometer from full resistance back to zero resistance. Repeat the action until you can clearly observe the voltage line movement.

Question 1. Describe the relationship of the flat trace line on the oscilloscope while the potentiometer is adjusted from maximum to minimum value. _____

_____

Question 2. What is the highest value of voltage indicated by the oscilloscope? _____ volts

Question 3. What is the lowest value of voltage indicated by the oscilloscope? _____ volts

Step 5. Turn off the power supply, disconnect the oscilloscope, and connect the red LED in parallel to the 4.7 kΩ resistor.

Step 6. Turn on the power supply to 12 volts dc once again.

Step 7. Slowly adjust the potentiometer from full resistance to zero resistance and then back to full resistance. Observe the performance of the LED.

Question 4. What effect does varying the potentiometer have on the LED connected to the transistor collector?

_____

_____

Step 8. Now you will observe the amount of voltage potential needed between the emitter and base for conduction through the collector-emitter circuit. Connect a voltmeter across the base (B) and emitter (E).

Step 9. Adjust the potentiometer until the LED goes completely out.

Step 10. Slowly adjust the potentiometer, stopping when the faintest amount of light is shown from the LED. Record the amount of voltage indicted by the voltmeter. _____ volts

Step 11. Adjust the potentiometer until the LED is at full brightness, and record the amount of voltage indicated on the voltmeter. _____ volts

 Question 5. What is the amount of voltage required between the emitter and base for the transistor to conduct current through the collector-emitter circuit? _____ to _____ volts

Step 12. Now remove the LED from the circuit, and connect the additional components to the circuit shown in the schematic that follows.

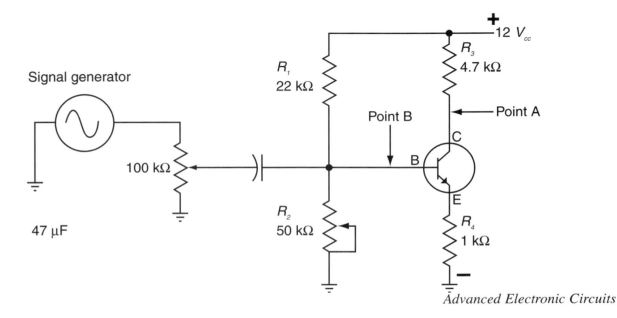

Step 13. Now you will observe the conditions required to amplify a sine wave signal injected into the base circuit. Adjust the oscilloscope to approximately: Time/Div. = 0.1 mS, Channel 1 = 2 volts, Channel 2 = 0.1 volts. Adjust the signal generator to a 2 kHz sine wave pattern. Be sure the probe for Channel 1 is connected to test point B and the probe for Channel 2 is connected to test point A.

Step 14. Adjust the 100 kΩ potentiometer until a 1.5 volt peak-to-peak sine wave is observed at point B. You may have to adjust the signal generator output voltage.

Step 15. Adjust the 50 kΩ potentiometer until a maximum undistorted value sine wave is observed. An undistorted sine wave at point A will be a similar shape to the sine wave at point B pattern. The wave should be symmetrical with no flat or sharp spots. You may need your instructor to assist you.

Step 16. Record the output peak to peak voltage at point A. _____ volts peak-to-peak. (Note: Do not forget to multiply by the appropriate Time/Div. factor on the oscilloscope.)

Question 6. What is the voltage gain of the transistor. Compare point A with point B. _____.

Question 7. Is the output sine wave at point A in phase or out of phase with point B? _____.

Step 17. Adjust the 50 kΩ potentiometer to the full resistance value and then its minimum value. Observe the shape of the output sine wave pattern as you adjust the 50 kΩ potentiometer.

Question 8. What effect does adjusting the value of the 50 kΩ potentiometer have on the base voltage?

_____

_____

Question 9. How is the output sine wave at point A affected by adjustments of the 50 kΩ potentiometer?

_____

_____

Question 10. Is the bias voltage at the base a critical item for proper operation of a distortion free amplifier?

_____

Question 11. If the transistor was used to amplify an audio sound, would the adjustment of the 50 kΩ

potentiometer to saturation affect the quality of the sound amplified? _____

Step 18. Readjust the 50 kΩ potentiometer until the input voltage at point B is approximately 1.5 volts peak-to-peak, and the output voltage is approximately 6 volts peak-to-peak with no distortion.

Step 19. Slowly increase the input signal frequency to the maximum value of the signal generator, and observe the wave shape of the output for distortion. You will have to adjust the time sweep of the oscilloscope as you increase the output frequency of the signal generator.

Question 12. Did the higher frequency distort the signal? _____

Step 20. Clear your work area. Properly store equipment and supplies.

A Light Activated Relay

Name _____ Score _____

Date _____ Class/Period/Instructor _____

Introduction

In this lab activity, you will assemble a simple light controlled relay. You will adjust the transistor circuit bias to de-energize the relay causing the lamp to come on. This circuit has many applications in industry.

Materials and Equipment

(1)—NPN switching transistor

silicon, $h_{FE} = 200$

$I_C = 800$ mA, $V_{CE} = 30$ V

power dissipation = 1.8 W

NPN

Collector Transistor pin
Base   identification
Emitter

Bottom view

(1)—photocell

(1)—12 Vdc relay, DPDT

(1)—silicon diode 1N4001, 50 V, 1 A

(1)—12 Vdc lamp and lamp holder

(1)—12 Vac power supply

(1)—oscilloscope

(1)—1-12 Vdc power supply

(1)—50 kΩ potentiometer

(1)—breadboard

Principles of Operation:

The photocell used in this activity is a cadmium sulfide type. The photocell is connected in series with the 50 kΩ potentiometer to form a voltage divider circuit for the base of the transistor. (See schematic in Step 2.) The photocell has a very high resistance in the absence of light and a lower resistance when exposed to bright light. The potentiometer can be adjusted to energize the transistor at the intensity of ambient light conditions. *Ambient light* is the surrounding room light.

When the photocell is covered to block out the light, the resistance value of the photocell increases. The increase of the resistance causes the bias voltage at the base to drop below turn on condition. The relay will de-energize, closing the normally closed contact connecting the lamp to the 12 Vac source. When ambient light is returned to the photocell, the relay will once again energize, turning the lamp off.

The purpose of the diode is to prevent a reverse kick of excessively high voltage (inductive reactance) caused by the collapse of the magnetic field of the relay coil. The circuit will operate correctly without the diode but it is standard practice to insert the diode to achieve a long product life.

Procedure

Step 1. Assemble the materials needed.

Step 2. Connect the circuit according to the schematic that follows. It is crucial that the diode be placed in the circuit with reverse bias. If the diode is placed in the circuit with forward bias, the diode and transistor will carry excessive current and will be damaged. Have your instructor check the wiring before you energize the circuit.

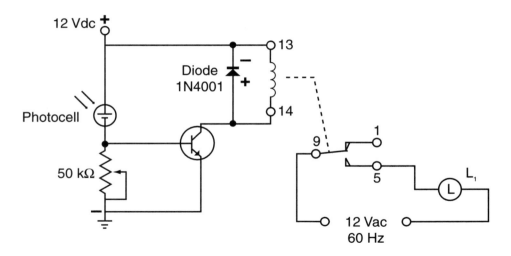

Some key points:
- Keep the dc voltage and ac voltage separate.
- Connect the diode with reverse bias.
- Make sure the lamp is connected to the normally closed contact of the relay.

Step 3. You will now adjust the potentiometer to the approximate cutoff value of the transistor bias. When adjusted properly, the lamp should be off and the relay energized.

Step 4. Move your hand across the photocell without touching it. Your hand should shade the light from striking the photocell and cause the relay to de-energize. If not, carefully adjust the 50 kΩ potentiometer until the desired effect is achieved.

Step 5. Connect the oscilloscope to the collector of the NPN transistor, and observe the fluctuation of the voltage level when the photocell is protected from the light source.

Question 1. How much did the voltage at the collector fluctuate when the photocell was exposed from light-to dark? _____ volts

Question 2. What other type of transducers could be used in place of the photocell?

_____

_____

_____

Question 3. What other type of relay could be used? _____

_____

Question 4. Could the relay be replaced by a transistor if dc was used to light the lamp? _____

Step 6. Clear your work area. Properly store equipment and supplies.

Transistor Gain

Name _____ **Score** _____

Date _____ **Class/Period/Instructor** _____

Introduction

In this lab activity, you will observe and calculate the gain of a transistor. It is important to understand the relationship between the base current and the collector current of a transistor. The relationship is called beta (β). Beta is the gain expressed as a numerical figure such as 50, 100, 200, etc. The gain of a transistor is expressed on a data specification sheet as h_{FE}. In this experiment, you will construct a simple, common emitter circuit to measure the current of the base, and compare it to the current through the load resistor.

Materials and Equipment

(1)—NPN transistor, MPS3904

(1)—50 kΩ potentiometer

(2)—1 kΩ resistors, 1/4 W

(1)—12 Vdc power supply

(2)—multimeters

(1)—breadboard

MPS3904

Collector 	Transistor pin
Base 	identification
Emitter

Bottom view

Procedure

Step 1. Collect all materials.

Step 2. Construct the circuit in the schematic that follows. Do not energize the circuit until your instructor has checked your work. Make sure that the 50 kΩ potentiometer is adjusted to its midrange.

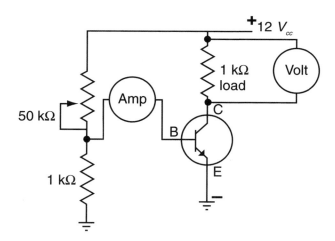

Key points:
- One multimeter is connected in series with the transistor base.
- One multimeter is connected in parallel with the load resistor.

Step 3. Adjust the 50 kΩ potentiometer until the base current equals approximately 10 microamps. This adjustment is difficult and requires a sensitive touch. You may have to settle for a value of plus or minus 10 percent. The results of the lab will still be the same.

Step 4. Record the readings of the base current below and the load resistor current. The load resistor current value is found by dividing the voltage drop across the load resistor by the value of the resistor in ohms. It is a simple application of Ohm's law. The base current is indicated by (Base I) and the load current is indicated by (Load I). Remember to express the current values in either microamp (μA) or milliamp (mA). Each cell below is an increment of 1 volt across the load.

Load V	1	2	3	4	5	6	7	8	9	10
Base I										
Load I										
Gain										

Step 5. After all the current values have been entered, calculate the gain. Simply divide the load current value by the base current value.

Question 1. What is the average gain of the transistor?_____

Question 2. Did the gain vary as the base current increased? _____

Question 3. What would the current value of the collector be equal to: the base or the load?_____

Question 4. Would the emitter current be higher, lower, or the same as the collector current? Explain your answer.

_____

_____

_____

Question 5. What would happen if the current rating of the collector was exceeded? _____

_____

_____

Step 6. Clear you work area. Properly store equipment and supplies.

Student Activity Sheet 18-7
The SCR

Name_____ **Score**_____

Date_____ **Class/Period/Instructor**_____

Introduction

In this laboratory activity, you will become familiar with the characteristics of the SCR. The SCR is classified as a thyristor rather than a transistor. Its actions closely resemble a diode that can be controlled by a gate. The SCR is used extensively in industry to control large current values. The SCR is often found in dc motor control systems because of their capability to handle large current values, some as high as 2000 amps. The SCR functions not only on dc but can also be used on ac circuits.

Materials and Equipment

(1)—12 Vdc supply

(1)—12 Vac supply

(1)—SCR

(1)—12 volt lamp

(1)—100 kΩ potentiometer

(1)—oscilloscope

(1)—breadboard

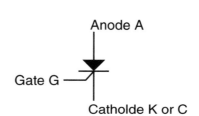

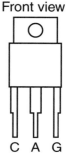

Procedure

Step 1. Assemble all required materials and calibrate the oscilloscope.

Step 2. Construct the circuit below. Have your instructor check your circuit before energizing.

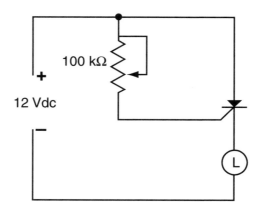

Key point:
* The SCR must be connected with the cathode to the negative and the anode to the positive side of the power supply. The gate also connects to the positive side of the power supply through the 100 kΩ potentiometer.

Step 3. Set the 100 kΩ potentiometer to full resistance value before energizing the circuit.

Step 4. Energize the circuit. If the potentiometer is set at full resistance, the lamp should not be lit when the circuit is energized.

Step 5. Slowly decrease the resistance of the 100 kΩ potentiometer. Closely observe the action of the lamp.

Step 6. Now, slowly increase the value of the potentiometer to full resistance again.

Step 7. Disconnect the wire from the 100 kΩ potentiometer and gate, and observe the lamp. It should continue to glow.

As you have demonstrated, once the SCR gate is energized, the current through the anode-cathode continues even when the gate is no longer energized. This is an important principle of the SCR operation.

Step 8. Reconnect the wire between the 100 kΩ potentiometer and the gate.

Step 9. De-energize the circuit, and connect the circuit to the 12 Vac supply.

Step 10. Once again, rotate the potentiometer from full resistance to zero resistance, and observe the lamp. This time, the brightness of the lamp coincides with the rotation of the potentiometer handle. The lamp became brightest when there was no resistance in the gate circuit and dark when full resistance was inserted in the gate circuit.

Step 11. Connect the oscilloscope between the SCR cathode and the lamp.

Step 12. Slowly rotate the potentiometer to control the lamp from full brightness to dark. Observe the scope pattern carefully.

Step 13. Draw the wave pattern below that corresponds with full brightness and half brightness.

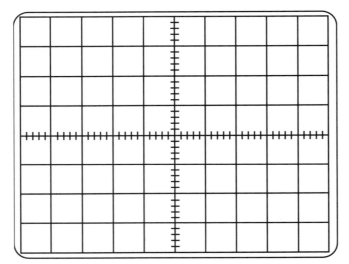

Full Brightness

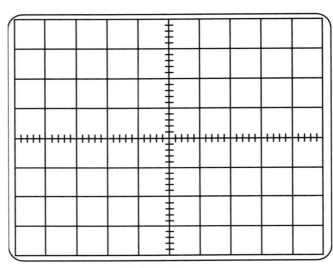

Half Brightness

Question 1. Based upon the scope patterns, what conditions must be present for the SCR to conduct? The answer should reference the gate polarity in relation to the conduction through the SCR.

_____

_____

_____

Question 2. Could an SCR be used as a rectifier? Explain? _____

Step 14. Clear your work area. Properly store equipment and supplies.

DC Motor Speed Control

Name_____ **Score**_____

Date_____ **Class/Period/Instructor**_____

Introduction

In this lab activity, you will build and operate a dc motor speed control using an SCR. You will apply your knowledge of power supplies to this activity. When the circuit is completed you will run the motor through various rpm while observing the sine wave pattern applied at the SCR cathode. You will also see the reason a large capacitor is required across a dc motor with a commutator.

Materials and Equipment

(1)—120 Vac supply (input for transformer)

(1)—9 Vdc motor

(1)—SCR

(1)—120 V to 6 V 1.2 amp center tap transformer

(1)—germanium diode or another (1N4001)

(1)—1N4001 diode

(1)—200 μF capacitor, 50 volt

(1)—10 μF capacitor, 50 volt

(1)—100 kΩ potentiometer

(1)—10 kΩ resistor, 1/4 W

(1)—1 kΩ resistor, 1/4 W

(1)—oscilloscope

Procedure

Step 1. Collect all materials, and calibrate the oscilloscope.

Step 2. Construct the circuit in the schematic that follows. Have your instructor check it before you energize the circuit.

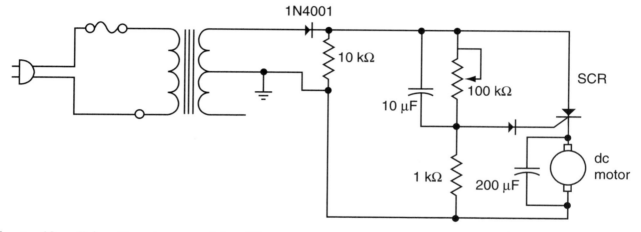

Key points:
- Pay particular attention to the polarity of the diodes.
- The gate diode works best as a germanium type rather than silicon.

Step 3. Energize the circuit, and rotate the potentiometer until the motor is rotating. See your instructor if the motor fails to run.

Step 4. Connect the oscilloscope to the SCR cathode. Observe the waveform while the motor is running.

Step 5. De-energize the circuit, and remove the 200 µF capacitor.

Step 6. Re-energize the circuit, and again, observe the oscilloscope wave pattern. Notice all the voltage spikes in the wave pattern. The spikes are caused by the commutator action of the dc motor. Many dc motor installations require capacitors mounted across the brushes to eliminate radio interference.

Step 7. *Challenge project*

Many control circuits in industry not only control the speed of a motor, but also the direction of rotation. Below, draw a circuit that will control the speed of the dc motor used in this project, and add components so that the motor will reverse direction. You may need to review some of the earlier projects dealing with motor control and relays. You can use multiple SCRs or relays.

Step 8. Clear your work area. Properly store equipment and supplies.

Integrated Circuits

Student Activity Sheet 19-1

Review

Name_____ **Score**_____

Date_____ **Class/Period/Instructor**_____

Complete the statements below by filling in the missing word or words.

1. The four devices commonly found on an integrated chip are _____, _____, _____, and _____.

1. _____

_____

_____

_____

2. After an integrated circuit is designed and drawn out, it is _____ reduced.

2. _____

3. Another name for the silicon wafer is the _____ layer.

3. _____

4. _____ is a growth of one crystal on the surface of another.

4. _____

5. There are two basic types of ICs, _____, and _____.

5. _____

_____

6. The 741 op amp is a(n) _____-_____ amplifier.

6. _____

7. Pin numbering locations are in relation to a(n) _____ point such as a(n) _____ on the chip.

7. _____

_____

8. The gain of an op amp is determined by the relationship of the _____ resistor and the _____ resistor.

8. _____

_____

9. The 555 IC is used for _____.

9. _____

10. The two main modes of operation for the 555 are _____ and _____.

10. _____

_____

The 741 Operational Amplifier

Name_____ **Score**_____

Date_____ **Class/Period/Instructor**_____

Introduction

In this lab activity, you will experiment with a 741 op amp. The 741 op amp is widely used in industry because of its simple design and excellent gain. It is much easier to use than the traditional transistor. It requires very few additional components to be utilized.

The two major uses of the op amp are as an inverting amplifier and as a noninverting amplifier. Pins 7 and 4 are the power supply connections. Pin 6 is the output of the amplifier, and pins 3 and 2 are the inputs of the amplifier. When a signal is applied to pin 2, the output will be the opposite polarity of the input. When a signal is applied to pin 3, the output will match the input polarity.

Materials and Equipment

(1)—741 op amp

(1)—dual voltage power supply or two nine-volt batteries

(1)—1 kΩ resistor, 1/4 W (brown, black, red)

(2)—2.2 kΩ resistors, 1/4 W (red, red, red)

(1)—10 kΩ resistor, 1/4 W (brown, black, orange)

(1)—22 kΩ resistor, 1/4 W (red, red, orange)

(1)—33 kΩ resistor, 1/4 W (orange, orange, orange)

(1)—100 kΩ resistor, 1/4 W (brown, black, yellow)

(1)—50 kΩ potentiometer

(1)—breadboard

(1)—dual-trace oscilloscope

Procedure

Step 1. Assemble all materials, and calibrate the oscilloscope (both channels).

Step 2. Construct the circuit that follows, but do not connect the power source to the circuit until your instructor checks it for accuracy. Insert the 1 kΩ resistor for R_1.

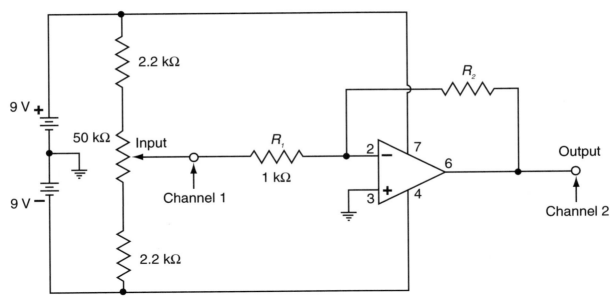

Key points:
- Note that the battery power supply is in series and a ground is established between them.
- R_2 is the feedback resistor that controls the amount of gain.

Step 3. Adjust the potentiometer to midrange, and turn on the power supply. If you are using nine-volt batteries, connect the ground first. Connect the negative and positive leads next.

Step 4. Connect Channel 1 of the dual trace oscilloscope to the input of the op amp and Channel 2 to the output of the op amp.

Step 5. Adjust the 50 kΩ potentiometer until there is a positive 0.05 volts indicated at the Channel 1 input. The potentiometer will cause either a negative or positive voltage at the input, depending on which way the potentiometer is rotated. Be sure there is a positive 0.05 volts at the input and not a negative 0.05 volts.

Step 6. Read and record the voltage and polarity of the voltage at channel 2 output, and record in the chart below. Replace R_2 resistor with each value indicated in the chart below, and then record the output voltage at Channel 2.

R_2	Input (V)	Output (V)	Gain
10 kΩ			
22 kΩ			
33 kΩ			
100 kΩ			

Step 7. Calculate the gain for each reading in the chart above.

Question 1. What is the relationship between the gain of the amplifier and the resistance values at R_1 and R_2?

_____

_____

Question 2. What is the relationship of the polarity of the input and the output voltage? _____

_____

_____

Step 8. Adjust the potentiometer to have a negative 0.05 V input. (The input was previously positive.) This will verify the conclusion to question 2.

Step 9. Remove the resistor at R_2, and leave it as an open circuit. Read the voltage output and record here. Output voltage when R_2 is equal to infinity = _____ volts

Question 3. What do you determine to be the maximum output voltage for this amplifier circuit? Read and record. Maximum output = _____ volts

Step 10. Disconnect the power, and reconnect the circuit to create the noninverting amplifier in the circuit schematic that follows.

Key points:

- Most of the components remain in the same position.
- The input from the potentiometer now connects to pin 3 and the R_1 and R_2 resistors connect to pin 2.
- The potentiometer connects directly to pin 3 with no resistance between.

Step 11. Reconnect the power supply.

Step 12. Adjust the potentiometer for an input of a positive 0.2 volts.

Step 13. Read and record the output voltage here. Output = _____ volts

Step 14. Slowly rotate the potentiometer from a positive one-volt input to a negative one-volt input while observing the oscilloscope.

Question 4. What is the relationship of the input voltage polarity and the output voltage polarity of the noninverting op amp? _____

_____

_____

Step 15. Clear your work area. Properly store equipment and supplies.

The 555 Timer

Name _____ **Score** _____

Date _____ **Class/Period/Instructor** _____

Introduction

In this lab activity, you will become familiar with the most common integrated timer in the world, the 555 timer. The 555 timer is simple to use in timing applications and requires very few additional components. The 555 timer is used mainly for two timing states, monostable and astable. Monostable is one complete timing cycle, a simple on/off operation for a duration of time. Astable is a repeating timing cycle, it will turn on and off repeatedly until the timing circuit is interrupted.

The pin arrangement is illustrated to the right of the materials and equipment list. Pins 1 and 8 are power supply connections. The output signal is taken from pin 3. The timing cycle is started by energizing pin 2, the trigger. Pin 4, the reset, is used to start the timing cycle again or to interrupt it. Pin 5 provides a means of controlling the timer turn on voltage level. Pin 5 is usually not required in most timing circuits and will simply be connected to a 0.01 µF capacitor. The timing cycle is controlled by pins 6 and 7.

Materials and Equipment

(1) — 0-12 Vdc power supply

(1) — 555 timer

(2) — 100 kΩ resistors, 1/4 W (brown, black, yellow)

(1) — 470 kΩ resistor, 1/4 W (yellow, violet, yellow)

(1) — 1 MΩ resistor, 1/4 W (brown, black, green)

(1) — 10 MΩ resistor, 1/4 W (brown, black, blue)

(1) — 0.01 µF capacitor, 50 V

(1) — 1 µF capacitor, 50 V

(1) — 10 µF capacitor, 50 V

(2) — LEDs

(1) — push-button switch, normally open

(1) — watch or clock

(1) — oscilloscope

(1) — breadboard

```
Ground 1 |          |  8 V+
Trigger 2 |    555   |  7 Discharge
Output  3 |          |  6 Threshold
Reset   4 |          |  5 Control Voltage
```

Procedure

Step 1. Gather all materials and calibrate the oscilloscope.

Step 2. Construct the schematic that follows. Have your instructor inspect it before you energize the circuit. The R_t resistor is 470 kΩ.

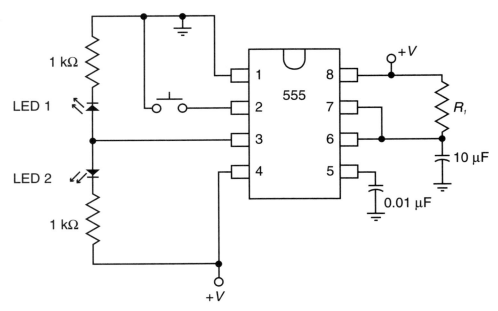

Key points:
- Carefully check the LEDs for proper bias.
- Check the polarity of the capacitors unless you are using a nonpolarized type.
- Be sure the push-button switch is a normally open type.

Step 3. Energize the circuit using 12 volts dc, and observe the two LEDs. One LED should be on and the other off. If not, check to see if one or both are reverse biased.

Step 4. Close the push-button switch and observe the LEDs. There should be a five second time delay (approximately) for this circuit based on the 470 kΩ resistor and the 10 μF capacitor. Timing for the 555 timer is based on the charge discharge time of R_1, and the 10 μF capacitor. The formula for computing the time is below.

$$\text{Time delay} = 1.1 \ (R_1 \times C)$$
$$\text{Time delay} = 1.1 \ (470{,}000 \times 0.00001)$$
$$\text{Time delay} = 5.17 \text{ seconds}$$

Step 5. Following is a chart with certain values of resistors and capacitors. Connect each combination to the 555 timer circuit, and measure the time delay with a watch. Also, calculate the expected time delay using the formula provided.

Resistor value	Capacitor value	Calculated time in seconds	Actual time in seconds
100 kΩ	10 μF		
470 kΩ	10 μF		
1 MΩ	10 μF		
10 MΩ	10 μF		
1 MΩ	1 μF		
10 MΩ	1 μF		

Advanced Electronic Circuits

Step 6. Compare the calculated time period to the actual time period.

Question 1. How do the calculated time periods compare to the actual measured time periods?

_____

_____

Question 2. Based on your knowledge of electronics learned thus far, what factors might affect the time accuracy of the circuit?_____

_____

_____

Step 7. Now you will determine if supply voltage will affect the 555 timer. Connect the 10 μF capacitor and the 470 kΩ resistor to the 555 timer circuit.

Step 8. The operating voltage range for the 555 is approximately 15 volts to 5 volts. Try the timing circuit at various voltage levels between 15 and 5 volts to see the effect of the voltage on the timing.

Question 3. How does supply voltage affect the operation of the 555 timer?_____

_____

_____

Step 9. Turn off the power supply, and reconstruct the circuit as indicated in the next schematic. This is an astable operation timer circuit. R_1 and R_2 are equal to 10 kΩ and C_1 is equal to 1 μF.

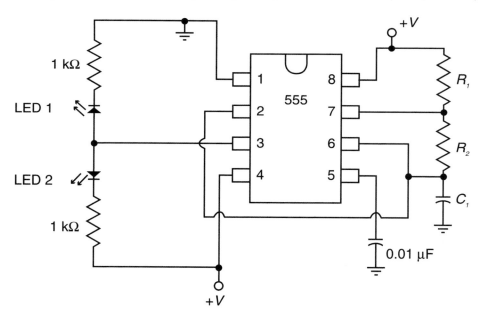

Key points:
- R_1 and R_2 must be in series with C_1.
- The push-button switch has been removed and pin 2 is now connected to the charge/discharge circuit of the resistor capacitor circuit.
- R_1 and R_2 must be in series with C_1.

Step 10. Connect the oscilloscope to pin 3 and observe the timing pulse.

Question 4. Is the wave pattern a sine wave, square wave, or triangular wave shape?_____

Step 11. Now connect the 100 kΩ resistor in the R_1 location, and leave the 10 kΩ resistor at the R_2 location. Observe the output pulse pattern on the oscilloscope.

Step 12. Reverse the location of the 100 kΩ and 10 kΩ resistor so that R_1 is now equal to 10 kΩ and R_2 is equal to 100 kΩ, and then observe the pulse output on the oscilloscope once again.

Question 5. Which resistor controls the time duration of the high, or positive, pulse and which resistor controls the low, or ground, pulse time duration at pin 3?

$$R_1 \text{ controls} \underline{\hspace{3cm}}$$

$$R_2 \text{ controls} \underline{\hspace{3cm}}$$

The formula for determining the frequency of the astable timer output is as follows.

$$f = \frac{1.44}{(R_1 + 2R_2) \times C}$$

Question 6. What would the frequency of the output of a 555 timer that used a 100 kΩ resistor for R_1 and a 1 MΩ resistor for R_2 connected to a 0.1 µF capacitor? Output frequency = \underline{\hspace{2cm}} Hz

Step 13. Clear your work area. Properly store equipment and supplies.

Student Activity Sheet 20-1

Review

Name_____ **Score**_____

Date_____ **Class/Period/Instructor**_____

Complete the statements below by filling in the missing word or words.

1. A potentiometer is an example of an analog circuit, while an SPST switch is an example of a(n) _____ circuit.

   1._____

2. The numbers used for binary counting are _____ and _____.

   2._____

   _____

3. A digital circuit has only two possible circuit conditions, it is either _____ or _____.

   3._____

   _____

4. The binary number 101 represents the decimal number _____.

   4._____

5. The binary number _____ represents the decimal number 8.

   5._____

6. A 4.2 volt signal would represent a valid logic _____ while a 1.2 volt signal would represent a valid logic _____.

   6._____

   _____

7. The term *bit* represents _____ _____.

   7._____

8. A *nibble* is equal to _____ bits of binary information.

   8._____

9. A *byte* is equal to _____ nibbles.

   9._____

10. If a two-input AND gate has only one input high, the output will be _____.

   10._____

11. If a two-input OR gate has one input high, the output will be _____.

   11._____

12. A NAND gate is composed of a(n) _____ gate and a(n) _____ gate.

   12._____

   _____

13. CMOS and TTL are examples of logic _____.

   13._____

14. A CMOS logic device is easily damaged by _____.

   14._____

15. A common device used for testing and evaluating digital circuits is the _____ _____.

   15._____

(Continued)

16. When dealing with flip-flops, R stands for _____ while S stands for _____.

16. _____

_____

17. The two outputs of a flip-flop are _____, and _____.

17. _____

_____

18. Binary counters can be constructed from _____-_____.

18. _____

19. A(n) _____ counter counts to ten.

19. _____

20. A(n) _____ counter is often used to divide by ten.

20. _____

Review–Part II

Name_____ **Score**_____

Date_____ **Class/Period/Instructor**_____

Match the symbols on the right with the word on the left that represents the symbol by putting the corresponding letter in the blank at the left.

1. _____ AND

2. _____ OR

3. _____ NAND

4. _____ NOR

5. _____ XOR

6. _____ XNOR

7. _____ NOT

A. ⟯‾)‾ ?   E. ⟯‾)o‾ ?

B. ⟯‾) ‾ ?   F. ▷o‾ ?

C. ⟯⟯‾)‾ ?   G. ⟯⟯‾)o‾ ?

D. ⟯‾)o‾ ?

Determine if the output is high or low for the logic circuits and devices below and then indicate in the space H for High and L for Low.

8. _____   H / L → ?

9. _____   H / L → ?

10. _____   H → ?

11. _____   H / L → ?

12. _____   H / L → ?

13. _____   H / H / L → ?

(Continued)

14. ______

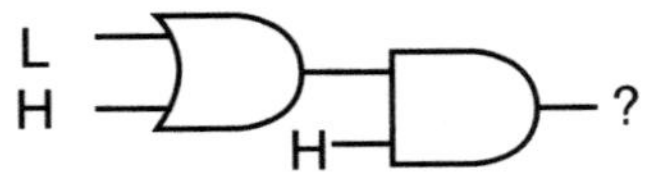

15. ______

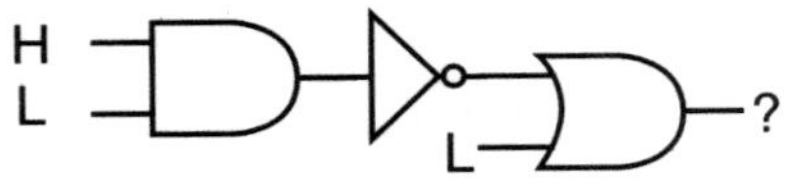

16. ______

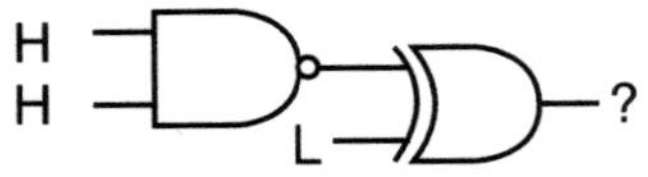

17. ______

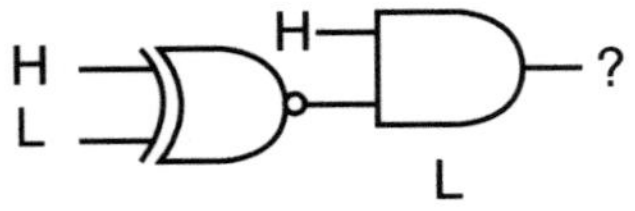

18. ______

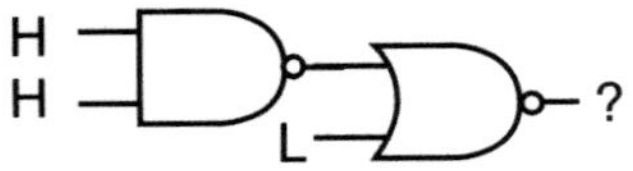

The AND, OR, and NOT Gates

Name_____ Score_____

Date_____ Class/Period/Instructor_____

Introduction

In this lab activity, you will become familiar with the AND and OR gates and a common inverter. These are the basic building blocks of digital electronics. Digital circuits revolutionized the electronics industry and are the basis of computer operation. Remember a digital basic—digital systems are either on or off. The high state of a digital circuit means there is a voltage present, usually +5 volts. The high state is expressed with a one (1) or the letter (H). The low state means the circuit is off, and is usually zero volts. It is expressed as a zero (0) or the letter (L).

Special Precautions for Handling CMOS: You must observe certain precautions when handling CMOS (Complimentary Metal-Oxide Semiconductor) chips, which can be easily damaged by static electricity.

1. Avoid touching the pins with your fingers. You could discharge static to the chip.

2. Always store the chip in a conductive foam or aluminum foil.

3. There are two pins for supply power on the chip. Always connect the ground first. Next connect the positive power supply pin before connecting power to any gate input or output pins. Energizing a gate pin without supply power could easily damage the chip.

4. Never disconnect the supply voltage pin while power is applied to any input gate pins.

5. Unused pins of a gate being used should be grounded. Pins left unconnected can cause erratic behavior. Your hands near the pins and chip may exhibit a static field of sufficient strength to cause the gate to appear as a high condition.

Materials and Equipment

(1)—5-volt power supply

(1)—7432 quad 2-input OR gate

(1)—7408 quad 2-input AND gate

(1)—7404 hex inverter

(1)—red LED

(1)—1 kΩ resistor, 1/4 W (brown, black, red)

(1)—5 kΩ potentiometer

(4)—SPDT switches

(1)—breadboard

(1)—logic probe

Procedure

Step 1. Gather all necessary materials. Be careful when handling CMOS devices. Observe recommended precautions.

Step 2. Connect the circuit shown in the schematic that follows. On the right is the gate arrangement on the chip. Take special note of the location of pin 7 (ground) and pin 14 (+5 volts). Remember to connect the ground first, and then the power supply pin before connecting inputs or outputs of gates. Pins 1 and 2 are the inputs to the AND gate, while pin 3 is the output. Connect one SPDT switch to pin 1 and the other switch to pin 2. The LED and 1 kΩ resistor are connected between pin 3 and ground. Once the circuit is connected, have your instructor check it for accuracy.

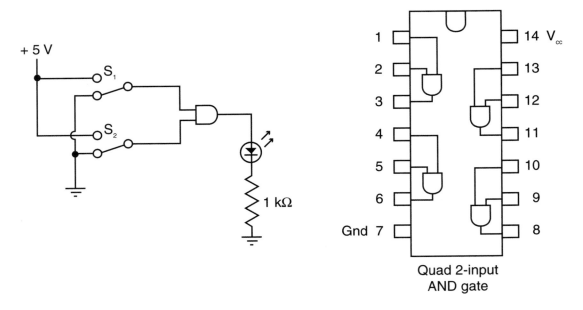

Quad 2-input
AND gate

Step 3. Now, you will complete the truth table for the AND gate. Set the SPDT switches for each condition in the truth table. When the switch is supplying +5 volts to the pin it is considered high, and when the switch is connecting the pin to ground, it is considered low. When pin 3 is high, the LED will be lit.

AND Gate Truth Table

Switch 1	Switch 2	Pin 3
Low	Low	
High	Low	
Low	High	
High	High	

Question 1. What condition is needed for the LED to light?_____

_____

Step 4. Draw the equivalent of an AND circuit using two single pole switches and a lamp connected to a 5-volt source.

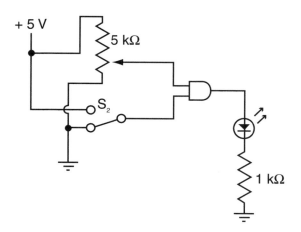

Step 5. Digital circuits have a high and low condition. A high condition is +5 volts, and a low condition is 0 volts, or ground. The actual voltage range at which digital devices will work varies from the 0 low level to +5 volt level. A digital gate can go to low at a voltage level of 0 to 0.8 volts, while a valid high condition can vary from 2 to 5 volts. To see this condition for yourself, wire the circuit that follows.

Replace switch 1 (S_1) with a 5 kΩ potentiometer wired according to the schematic. Connect a voltmeter to measure the voltage at pin 1.

Step 6. Vary the 5 kΩ potentiometer to determine the voltage range of the high condition for the AND gate. Switch 1 should be in the high condition mode.

Question 2. What minimum voltage level was needed for a condition high (LED is lit). _____ volts

Step 7. Properly store the AND gate by placing it in conductive foam.

Step 8. Connect the OR gate according to the schematic that follows.

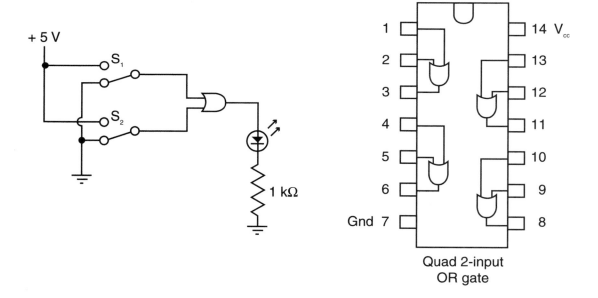

Quad 2-input
OR gate

Step 9. Complete the truth table for the OR gate by varying the input switches to simulate high and low conditions.

OR Gate Truth Table

Switch 1	Switch 2	Pin 3
Low	Low	
High	Low	
Low	High	
High	High	

Question 3. What condition is needed for the LED to light?_____

_____

Question 4. Draw an equivalent of an OR gate using two single pole switches and a lamp connected to a 5-volt source.

Step 10. Remove and properly store the OR gate in conductive foam.

Step 11. Insert the hex inverter into the breadboard, and construct the circuit that follows. You can use S_1 and remove S_2 to complete the inverter circuit. After you have completely constructed the circuit, vary the input at pin 1 from high to low and observe the LED.

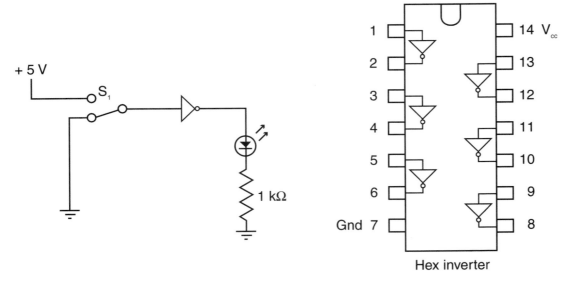

Hex inverter

Advanced Electronic Circuits

Question 5. What is the relationship of the input to the output of an inverter?

_____

_____

Question 6. The chip marked 7408 is a(n) _____ gate, while the chip marked 7432 is a(n) _____ gate.

Question 7. Pin number _____ was used as a ground on the 7408 while pin number _____ was the positive voltage connection that supplied power to the chip.

Question 8. In general, a low condition is equal to _____ volts, while a high condition is equal to _____ volts.

Step 12. Clear your work area. Properly store equipment and supplies. Use conductive foam for storing CMOS devices.

The NAND, NOR, and Exclusive OR Gates

Name_____ **Score**_____

Date_____ **Class/Period/Instructor**_____

Introduction

In this laboratory activity, you will become familiar with the basic operation of the NOR, NAND, and exclusive OR logic gates. When completed, you will have observed the operation of all basic logic gates, and you will be ready to move on to more complicated devices used in digital systems.

Materials and Equipment

(1)—5-volt power supply

(1)—4011 quad 2-input NAND gate

(1)—4001 quad 2-input NOR gate

(1)—74HCT86 quad 2-input exclusive OR gate

(1)—red LED

(1)—1 kΩ resistor, 1/4 W (brown, black, red)

(4)—SPST switches

(1)—breadboard

(1)—logic probe

Procedure

Step 1. Review the precautions for handling CMOS materials. (See Student Activity Sheet 20-3.)

Step 2. Gather all necessary materials.

Step 3. Connect the circuit below. Take special note of pin 7 (Gnd) and pin 14 (V_{cc}). Connect pins 7 and 14 first, before any power is applied to the circuit. Pins 1 and 2 are the inputs, while pin 3 is the output. Have your instructor check the circuit before energizing.

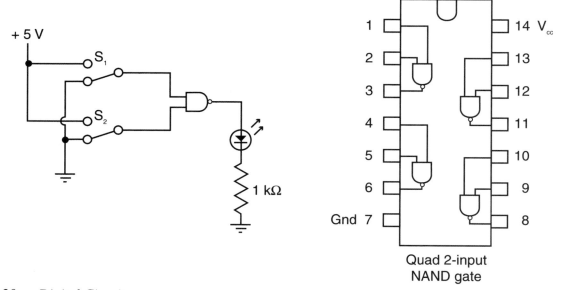

Quad 2-input
NAND gate

Step 4. Now complete the truth table below for the circuit. Use the switches S_1 and S_2 to set up the input conditions. Record the results (output) in the NAND truth table.

Switch 1	Switch 2	Pin 3
Low	Low	
High	Low	
Low	High	
High	High	

Question 1. What condition is needed for the LED to light?_____

_____

Step 5. Now insert the NOR gate in place of the NAND gate so that it matches the circuit that follows. Properly store the NAND.

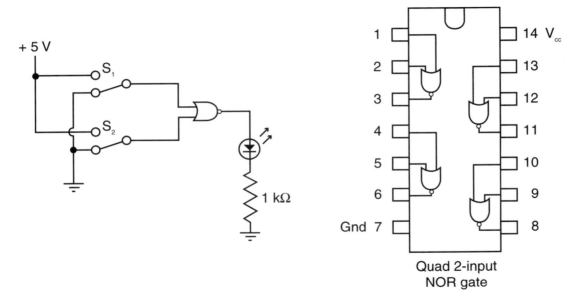

Quad 2-input
NOR gate

Step 6. Use the switches S_1 and S_2 to simulate all the conditions for the following NOR truth table. Record your results by completing the NOR gate truth table.

Switch 1	Switch 2	Pin 3
Low	Low	
High	Low	
Low	High	
High	High	

Question 2. What condition is needed for the LED connected to pin 3 of the NOR gate to light?

_____

Step 7. Replace the NOR gate with the exclusive OR gate. See the schematic that follows.

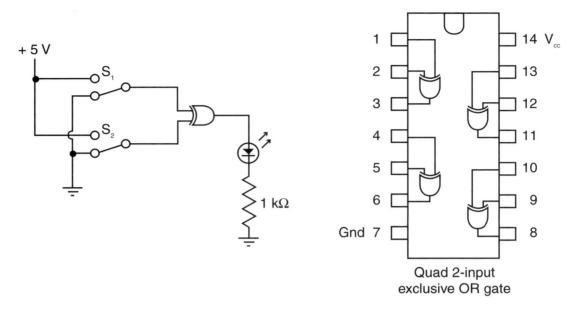

Quad 2-input
exclusive OR gate

Step 8. Complete the exclusive OR truth table for the conditions specified by S_1 and S_2. Record the results in the exclusive OR truth table.

Switch 1	Switch 2	Pin 3
Low	Low	
High	Low	
Low	High	
High	High	

Question 3. What conditions caused the LED connected to the output to light? _____

_____

Question 4. Compare and contrast the AND gate truth table from the last laboratory activity to the NAND truth table in this lab activity._____

_____

_____

Question 5. Compare and contrast the OR gate truth table from the last lab activity to the NOR gate truth table from this lab activity._____

_____

_____

Question 6. What is different between the OR gate and the exclusive OR gate? _____

_____

_____

Question 7. Could you make an exclusive NOR gate, and if so, how? _____

_____

_____

Question 8. Create a truth table for a two input exclusive NOR gate in the space that follows.

Step 9. Clear your work area. Properly store equipment and supplies. Use conductive foam for storing CMOS devices.

Multiple Input AND Gates

Name_____ **Score**_____

Date_____ **Class/Period/Instructor**_____

Introduction

In this lab activity, you will learn how to convert a 2-input logic device into a 3- or more input logic device. By comparing the technique used for the multiple input AND, you can also create multiple inputs for any standard logic gate device.

Materials and Equipment

(1)—quad 2-input AND gate (#7408)

(1)—5-volt power supply

(4)—SPDT switches

(1)—red LED

(1)—1 kΩ resistor, 1/4 W (brown, black, red)

(1)—breadboard

(1)—logic probe

Procedure

Step 1. Assemble all materials needed.

Step 2. Construct the circuit in the schematic that follows. Do not energize until the instructor has checked your work. Be sure to follow all precautions when using CMOS devices.

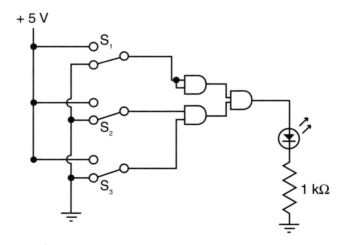

Key Point:
- Inputs of one AND gate are connected to form one single input device.

Step 3. In the space provided below, design a truth table for all possible combinations of input switches. Then, fill in the output results based upon the results of the lab.

Question 1. Could a quad 2-input OR gate be designed as a 3-input device?_____

Question 2. Design and draw a 4-input AND gate in the space that follows. Use only a total of three AND gates from the same 2-input quad AND gate.

Question 3. Design and draw a 4-input OR gate below. Use only a total of three OR gates from a quad 2-input OR chip.

Question 4. Design and draw a truth table of the circuit you just created. The table should depict all possible combinations of high and low conditions at the inputs.

Step 4. Clear your work area. Properly store equipment and supplies. Use conductive foam for storing CMOS devices.

NAND Gated R-S Flip-Flop

Name_____ **Score**_____

Date_____ **Class/Period/Instructor**_____

Introduction

In this lab activity, you will construct an R-S flip-flop from a 2-input quad NAND gate. After constructing the R-S flip-flop, you will observe and record the results of the R-S flip-flop device under various conditions.

The R and S are inputs that are generally referred to as set and reset. Take note of the bubble on the input. The bubble means that the outputs will be inverted. When S is high, Q will be low, and when S is low, Q will be high. Q and $\overline{Q}$ are complementary. They have opposite outputs, except in a prohibited state. Flip-flops are used to construct sequence circuits and memory devices. These devices will be explored in later labs after you have mastered the basic concepts of various flip-flops.

Materials and Equipment

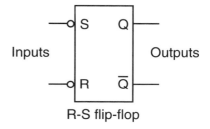

    (1)—4011, 2-input quad NAND gate

    (1)—5 Vdc power supply

    (1)—555 timer

    (2)—SPDT switches

    (1)—1 kΩ resistor, 1/4 W (brown, black, red)

    (2)—red LEDs

    (1)—breadboard

    (1)—logic probe

R-S flip-flop

Procedure

Step 1. Gather together all required materials.

Step 2. Construct the schematic that follows. Have your instructor check your project before energizing.

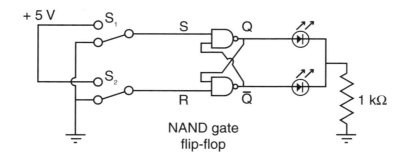

NAND gate
flip-flop

Key Point:

• The output of each NAND gate is connected to the input of the opposite NAND gate.

Step 3. Energize the circuit, +5 volts maximum. Complete the flip-flop truth table that follows. Match the switches to the input conditions given, even the prohibited condition. Record your findings.

S₁ (S)	S₂ (R)	Q	Q̄
Low	Low	XXXX	XXXX
Low	High		
High	Low		
High	High		

XXXX = Prohibited state — it is disallowed.

The condition (S_1 = low, S_2 = low) is a prohibited or disallowed state because it is a unpredictable state. The last condition (S_1 = high, S_2 = high) needs to be watched closely. Take note of the fact that this condition is dependent on the previous condition. It locks or latches the previous condition.

Question 1. A high at R and a low at S will cause a _____ at Q and a _____ at Q̄.

Question 2. A low at R and a high at S will cause a _____ at Q and a _____ at Q̄.

Question 3. A high at R and a low at S followed by a high at S will cause a _____ at Q and a _____ at Q̄.

Question 4. A low at R and a high at S followed by a high at R will cause a _____ at Q and a _____ at Q̄.

Question 5. Which condition acts like a latch, or a memory, of the previous state?

_____

_____

Step 4. Clear your work area. Properly store equipment and supplies. Use conductive foam for storing CMOS devices.

D Type Flip-Flop

Name_____ **Score**_____

Date_____ **Class/Period/Instructor**_____

Introduction

In this laboratory activity, you will experiment with a D type flip-flop. The D flip-flop derives its name "D" from the data, or delay, of the output condition. When data is input to the D type flip-flop, the output is delayed until a signal is present at D.

The S and R are set and reset or clear. They are inputs that act independently of other data inputs. They will either clear the output condition or save it. The C is the clock and provides the timing for the unit. Most applications require a timing or clock system to assure that the flip-flop is activated in sequence with other devices. The input data will not affect the output data at Q and $\overline{Q}$ until the C is enabled (pulsed by the clock circuit). There are two types of clock circuits, synchronous and asynchronous. Synchronous is in pace with the clock circuit, and asynchronous is out of pace with the clock circuit pulses. Study the diagram of the dual D flip-flop to the right of the materials and equipment list. The pin configuration matches only a 7408 type flip-flop. If another type of D flip-flop is substituted, the pins may not match.

Materials and Equipment

- (1)—7408 dual D flip-flop
- (1)—5 Vdc power supply
- (2)—red LEDs
- (1)—1 kΩ resistor, 1/4 W (brown, black, red)
- (4)—SPDT switches
- (1)—breadboard

Q – Normal output

$\overline{Q}$ – Complements Q

C – Clock

R – Reset

D – Data

S – Set

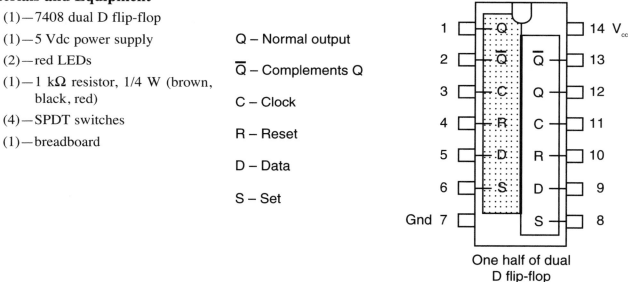

One half of dual
D flip-flop

Procedure

Step 1. Gather all required materials. Observe all precautions when handling CMOS materials.

Step 2. Construct the circuit below. Do not energize the circuit until your instructor has checked your work.

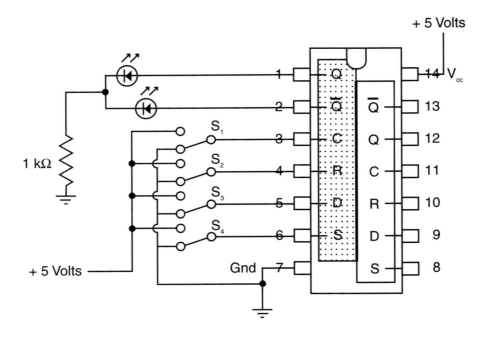

Key Points:

- Note the anode and cathode on the two LEDs.
- Each SPDT switch has a ground and a positive 5-volt source connected to it.
- Each switch is at the low or ground connection status.

Step 3. Set each switch to a low input condition. Check each input from each switch with a logic probe. The logic probe should be set in the CMOS condition. Each switch input should read as low.

Step 4. Toggle the reset switch (S_2) to the high and then back to the low input condition.

Question 1. What is the status of the Q and $\overline{Q}$ output LEDs? Record your findings as a high (lit) or low (not lit).

Q LED is _____

$\overline{Q}$ is _____

Step 5. Toggle the data input (S_3) to a high and then to a low condition.

Question 2. Did the output status of the LEDs change? _____

Step 6. Toggle the clock input (S_1) to a high, and then return it to the low condition.

Question 3. Did the output status of the LEDs change? _____

Step 7. Put the data input (S_3) to the high condition, and leave it on high.

Step 8. Now put the clock input (S_1) to a high condition, and closely observe the output of the LEDs.

Question 4. The output of Q and $\overline{Q}$ changed status only when what condition was present?

_____

_____

Step 9. Return data and clock to a low input condition and leave them as lows.

Step 10. Toggle reset (S_2) to a high and then low condition. Leave reset in the low condition. Q should be low, and $\overline{Q}$ should be in the high condition.

Step 11. Place Set (S_4) in the high condition, and try to change the output status of Q and $\overline{Q}$ using only the data (S_3) and clock (S_1) switches.

Question 5. Can the output status of Q and $\overline{Q}$ be changed when set is high? _____

Step 12. Change set (S_4) to a low condition, and try changing the status of Q and $\overline{Q}$ by manipulating (S_1) clock and (S_3) data switches.

Question 6. Fill in the missing words. (data, set, clock, reset)

The _____ input and the _____ input are used together to manipulate the output condition of Q and $\overline{Q}$. The data input controls the output status of Q and $\overline{Q}$ after the _____ has been toggled. The output status of Q and $\overline{Q}$ can be retained if _____ is in a high condition. With all switches in the low condition, the _____ can be toggled high and then back to low, which will return the outputs to the normal output condition, Q as a low and $\overline{Q}$ as a high.

Question 7. Why would the D type flip-flop make a good memory device?

_____

_____

Step 13. Clear your work area. Properly store equipment and supplies. Use conductive foam for storing CMOS devices.

The J-K Flip-Flop

Name_____ **Score**_____

Date_____ **Class/Period/Instructor**_____

Introduction

The last flip-flop device with which you will experiment is the J-K flip-flop. The J-K flip-flop is probably the most widely used flip-flop in digital electronics. It has all of the electrical qualities of the previous flip-flops, plus additional features which make it unique. The inputs are labeled J and K. When both inputs (J and K) are at a low, the flip-flop is in a hold, or set, condition. The output status of Q and $\overline{Q}$ are held steady. When both the J and K input are in the high condition, the outputs will toggle high and low in synch with the clock pulses. This feature is very useful in counter applications.

Look at the illustrations to the right of the materials and equipment list. The logic symbol for a single J-K flip-flop is drawn beside the pin diagram of the 7476 dual J-K flip-flop. The ground and power supply are commonly not drawn on in the logic symbol but are considered to be understood. Take note of the pin markings on the 7476. The 7476 consists of two J-K flip-flops. The first flip-flop is noted as J1, K1, Q1, and $\overline{Q}$1. The second flip-flop is marked J2, K2, Q2, and $\overline{Q}$2. There is one common ground and voltage supply connection. Each flip-flop has its own preset, clear, and clock pin.

Materials and Equipment

(1)—J-K flip-flop 7476

(1)—5 Vdc power supply

(1)—1 kΩ resistor, 1/4 W
  (brown, black, red)

(5)—SPDT switches

(2)—red LEDs

(1)—breadboard

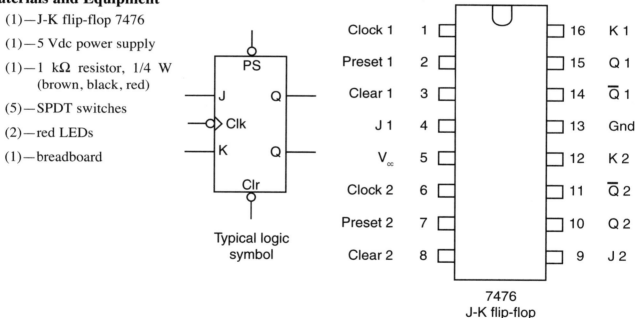

Typical logic
symbol

7476
J-K flip-flop

Procedure

Step 1. Assemble all materials required for the project.

Step 2. Construct the circuit in the schematic that follows. Have your instructor check your project before energizing.

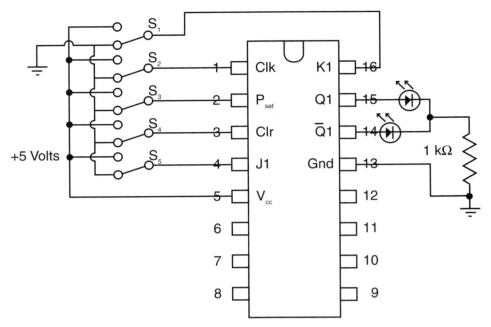

Key points:

Each switch has a ground and a +5 volts connected to it. The common on each switch ties to a pin to provide either a high or low signal to that pin.

Double-check the pin connections; you may wish to use different color wire to help keep track of the correct wiring.

Step 3. Have your instructor inspect your project before proceeding further.

Step 4. Energize the power supply, and then carefully follow the remaining steps.

Step 5. Make pin 4 (J2) high and pin 16 (K1) low.

Step 6. Set pin 3 (Clr) to high. Toggle pin 2 (P_{set}) high and then low and back to high.

Step 7. Now toggle the clock, pin 1, and observe the effect on Q and $\overline{Q}$ (LEDs at pin 15 and pin 14).

Question 1. What happens to Q and $\overline{Q}$ as the clock is pulsed from high to low and back to high?

_____

_____

Question 2. Toggle the clock from high to low several more times, and then report your observation.

_____

_____

Step 8. Now set the inputs J1 and K1 to the opposite condition, run the same test, and compare the results. Set pin 4 (J1) low and pin 16 (K1) low.

Step 9. Set the pin 3 (Clr) to high. Toggle pin 2 (P_{set}) high and then low and back to high.

Step 10. Now toggle the clock (pin 1) and observe the effect on Q and $\overline{Q}$ (LEDs at pin 15 and pin 14).

Question 3. What happens to Q and $\overline{Q}$ as the clock is pulsed from high to low and back to high?

_____

_____

Advanced Electronic Circuits

Question 4. Toggle the switch feeding the clock several more times to see the effect on the LEDs. Report the effect here. _____

_____

Step 11.   Now set both the J1 and K1 to high inputs.

Step 12.   Toggle the Clr and the P_{set} pins from high to low and then back to high.

Step 13.   Toggle the clock from high to low to high several times, and observe closely the effect on the LEDs.

Question 5. Describe the effect of the clock pulses on the LEDs at Q and $\overline{Q}$ when all pins (Clr, P_{set}, J1, and K1) are held in a high condition._____

_____

_____

Step 14.   Clear your work area. Properly store equipment and supplies.

J-K Flip-Flop Binary Counter

Name _____ **Score** _____

Date _____ **Class/Period/Instructor** _____

Introduction

In this lab activity, you will build a simple binary counter. You will first construct a 555 timer circuit to supply a consistent digital pulse and act as a bounceless switch for smooth operation. Two 7476 J-K flip-flops will be wired together to form a four-bit counter. The LEDs will light in progression from binary 0001 to 1111.

The project may look complex, but it is really quite simple to construct. Take note of the fact that all the pins marked as P_{set}, Clr, J, and K are tied to +5 volts. Each LED is connected to a Q. The secret to the progressive counting is the sequence of pulses through the clock circuits. See how the first clock, pin 1 on the top J-K flip-flop, receives its timing pulse from the 555 timer. After that, each Q LED feeds the clock of the next flip-flop. The 555 timer circuit was built in a previous lab. The erratic behavior of the last lab was a result of the switches used to pulse the circuit. The switches would occasionally bounce on contact. The use of the 555 timer as the source of the pulses will eliminate the bounce from the signal.

Materials and Equipment

(2)—J-K flip-flops 7476

(1)—5 Vdc power supply

(4)—1 kΩ resistors, 1/4 W (brown, black, red)

(2)—10 kΩ resistors, 1/4 W (brown, black, orange)

(1)—555 timer

(1)—47 μF capacitor

(1)—breadboard

Procedure

Step 1. Assemble all materials.

Step 2. With the power off, construct the circuit that follows.

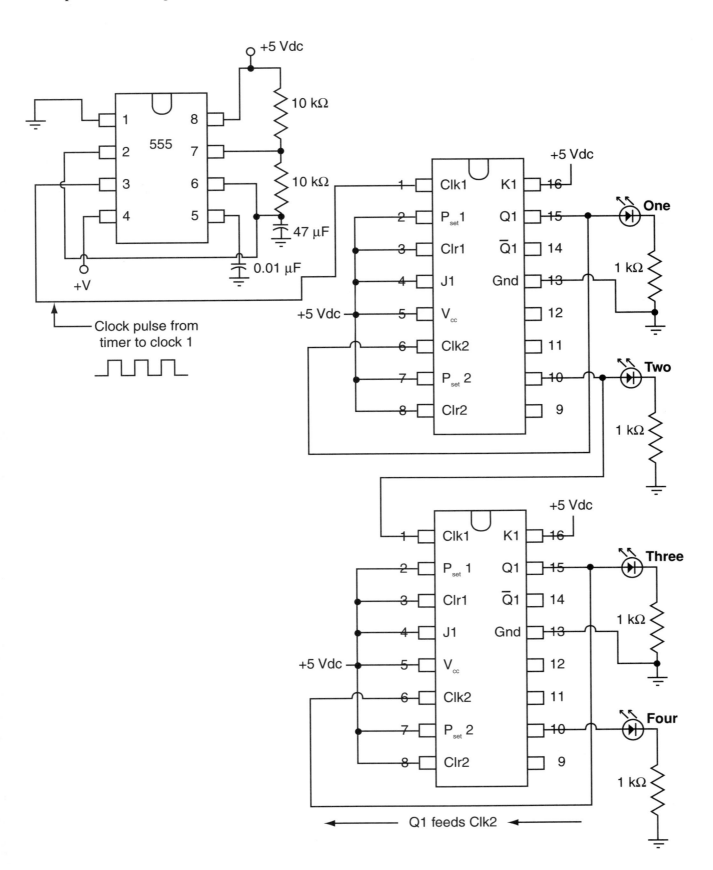

Step 3. Have your instructor check your work.

Step 4. Energize the circuit and closely observe the action of the LEDs.

Question 1. Using the following symbols, draw this lab circuit as a logic circuit.

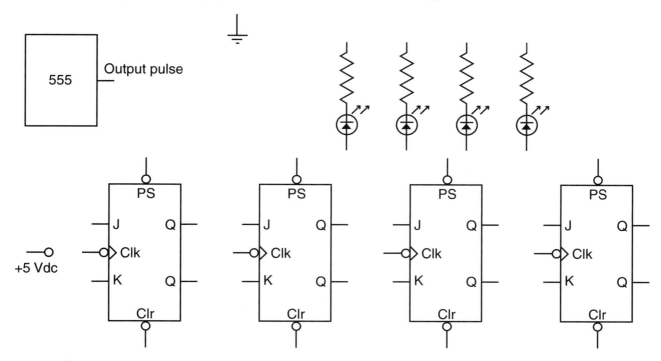

Step 5. Now turn off the power supply and reconnect the circuit as a down counter. Simply connect each LED to $\overline{Q}$ (pin 14) and each clock feed from $\overline{Q}$ instead of Q.

Step 6. Have your instructor inspect your work before progressing further.

Step 7. Energize the circuit and closely observe the sequence of the LEDs. If the LEDs blink too fast for you to see the sequence, increase the capacitor size. This will slow down the blinking rate.

Question 2. What is the maximum count that the four LEDs can represent as a decimal number? _____

Question 3. What is the limit to the number of flip-flops that can be used for this type of counter? Explain?

_____

_____

Question 4. Draw the pulse pattern of the Q outputs below.

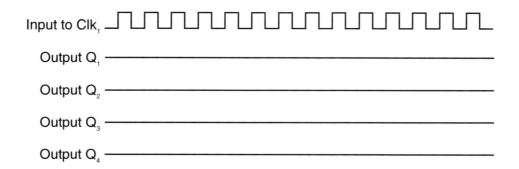

Step 8. Clear your work area. Properly store equipment and supplies.

Binary Ripple Counter

Name_____ **Score**_____

Date_____ **Class/Period/Instructor**_____

Introduction

In this lab activity, you will see a binary ripple counter that is formed into a single chip rather than using individual flip-flops. Since counting is such a basic electronic control system, it is only natural to construct a counter on a single chip. The 74HCT393 is a dual 4-bit binary ripple counter. This means that there are two counters constructed in the chip. The "4-bit" refers to the maximum possible binary number for each counter. "Ripple" counter refers to the style of count which you will see.

Look at the pin diagram below. Pin 1 and 2 are marked 1CP and 1MR. 1CP stands for clock pulse, and 1MR represents master reset for counter one. The Qs represent the outputs. As you can see, this style of ripple counter is much simpler than one constructed from individual flip-flops.

Materials and Equipment

    (1)—5 Vdc power supply

    (1)—74HCT393 ripple counter

    (1)—555 timer (from last project)

    (8)—LEDs

    (8)—1 kΩ resistors, 1/4 W (brown, black, red)

    (2)—10 kΩ resistors, 1/4 W (brown, black, orange)

    (1)—47 μF capacitor

    (1)—SPST switch

    (1)—logic probe

    (1)—breadboard

1 CP	1	14	V_{cc}
1 MR	2	13	2 CP
1 Q_0	3	12	2 MR
1 Q_1	4	11	2 Q_0
1Q_2	5	10	2 Q_1
1 Q_3	6	9	2 Q_2
Gnd	7	8	2 Q_3

74HCT393
Binary Ripple Counter

Procedure

Step 1. Assemble all materials.

Step 2. With the power off, construct the circuit in the schematic that follows. Have your instructor check the project before you energize the circuit.

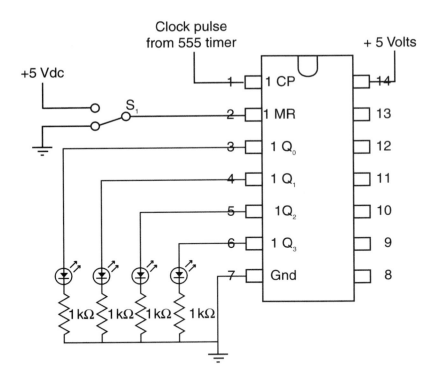

Key points:

- If the circuit fails to count try these items.
- Make sure that switch 1 is connected to a low condition.
- Be sure of the anode and cathode connections on the LEDs.
- Check the timer output pulse used for the clock circuit with either an LED or a logic probe.

Step 3. Energize the circuit. Set S_1 to a low condition and observe the light pattern. Be sure the 555 timer circuit is providing a pulse.

Question 1. Assuming that the first number represented by the binary counter 0000 equals zero, what is the highest decimal number represented by one side of the counter? Binary = _____, Decimal = _____

Complete a wiring diagram below to extend the range of the counter to an 8-bit counter. You will connect four LEDs to the 2Q side of the chip. You must solve how to drive the clock for the second half of the timer. Look back at the flip-flop circuits for a hint. The second half of the timer should start a clock pulse after the highest value (1111) has been reached by the first counter. After you have completed the drawing, have your instructor check it for accuracy.

Step 4. With the power off, construct the circuit according to your drawing. Have your instructor check it after you have completely wired it, but before you energize it.

Step 5. Energize the circuit and closely observe the light pattern displayed by the LEDs.

Question 2. What is the highest binary and decimal equivalent number displayed by the LEDs when using both sides of the counter to make a 8-bit counter? Binary = _____ , Decimal = _____

Question 3. Look carefully at the first wiring diagram of the counter and determine where you would connect a logic gate in this circuit to limit the count to only decimal ten. Also, determine what type of gate (AND, NAND, OR, NOR). Draw your solution schematic below in the space provided.

Which type of gate? _____

Question 4. Explain how your design would limit the count to only decimal ten, and would then repeat the process of counting to ten again.

_____

_____

Step 6. Clear your work area. Properly store equipment and supplies.

Displaying a Digital Pulse Rate

Name_____ **Score**_____

Date_____ **Class/Period/Instructor**_____

Introduction

In this lab activity, you will construct a circuit that will count digital pulses and display the count on a seven segment display. This is a basic circuit used in such items as frequency counters, clocks, timing device displays, speedometers, and many more electronic devices. Looking at the figure below, you can see the basic operation of the lab you will construct.

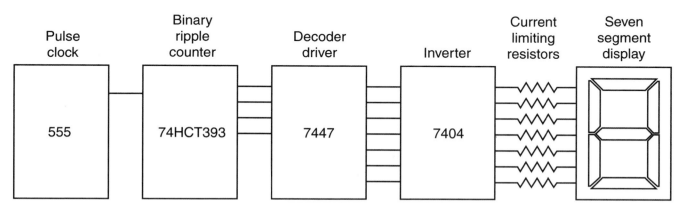

This circuit consists of a 555 timer to produce a steady pulse. A 74HCT393 ripple counter will count the number of pulses and convert the pulses to a binary output. The binary output from the ripple counter will be fed into a 7447 decoder/driver. The decoder/driver will decode the binary input from the ripple counter and convert it to an output (driver) that will drive the seven segment display. Between the 7447 decoder/driver and the seven segment display is a 7404 inverter. The inverter is necessary because the seven segment display has a common cathode (negative common connection). If the display had a common anode (positive common connection), the inverter would not be needed.

Let's take a closer look at the 7447 decoder/driver. The pin layout is shown to the right of the list of materials and equipment. Pin 8 is ground and pin 16 is connected to 5 volts dc. Pins 1, 2, 6, and 7 are the inputs. Each input is labeled with an uppercase letter (A, B, C, D). The outputs are from pin 9 through pin 15. The outputs are also labeled with letters, but the letters are lowercase to assist you in distinguishing between the inputs and the outputs. The outputs are labeled as (f, g, a, b, c, d, e). It is important to note that the outputs are not in sequential order. Pins 3, 4, 5 are used to control the outputs of the 7447 decoder/driver. They are connected to high and low conditions to blackout the display (make all segments dark) or to test the segments of the display (make each segment light).

Materials and Equipment

    (1)—5 Vdc power supply

    (1)—555 timer (saved from previous projects)

    (1)—binary ripple counter 74HCT393

    (1)—decoder/driver 7447

    (2)—hex inverters 7404

    (1)—LED seven segment display, common cathode

    (7)—390 W resistors, 1/4 W (orange, white, brown)

    (1)—breadboard

    (1)—logic probe

Chapter 20   Digital Circuits

Procedure

Step 1. Assemble all materials.

Step 2. Look at the partial wiring diagram that follows. The pulse from the 555 timer connects to the 1CP (Pin 1) on the binary ripple counter. The $1Q_0$ (Pin 3) connects to input A (pin 7) on the decoder/driver. The output a (pin 13) connects to the corresponding a (pin 14) on the seven segment display after going through the inverter gate (pin 1 and 2) on the hex inverter. Now complete the remaining connections by repeating this pattern. The binary ripple counter outputs Q_0, Q_1, Q_2, Q_3, connect to the corresponding A, B, C, D inputs of the decoder/driver. The decoder/driver outputs f, g, a, b, c, d, e connect to the corresponding letters on the seven segment display through an inverter. Indicate all ground and power (+5 Vdc) connections as needed in the drawing.

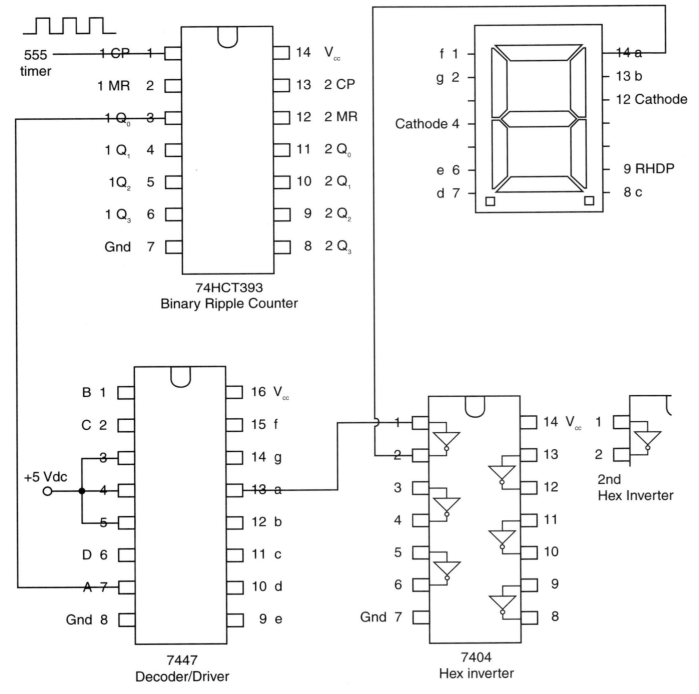

Step 3. Have your instructor check your drawing for accuracy before proceeding.

Step 4. With the power off, connect the circuit as indicated on the wiring diagram you made. Do not energize the circuit at this point.

Step 5. Have your instructor check the circuit you have wired before proceeding.

Step 6. Energize the circuit and closely watch the display. A series of numbers from 0 through 9 should be displayed, followed by some other shapes that resemble the letters C, U, and others. The display should then repeat again the series and continue as long as power is applied. If the display does not work, start tracing the circuit with the logic probe. Start at the clock pulse to be sure you have a signal. Then sequentially trace the signal from the timer to the display. You may have a loose connection or an inappropriate ground or power connection.

Question 1. How could this circuit be modified to eliminate the unwanted symbols displayed after the 9?

_____

_____

_____

_____

Question 2. How could this circuit be modified to display a count form 1 to 100? _____

_____

_____

_____

_____

Question 3. How could this circuit be modified to a digital clock unit displaying hours, minutes, and seconds?

_____

_____

_____

_____

Step 7. Connect pins 3, 4, 5 of the 7447 decoder/driver in combinations of highs (+5 Vdc) and lows (ground) to observe the effect on the output of the display unit. Run tests until you thoroughly understand the function of the lamp test (pin 3), RB out (pin 4), and RB in (pin 5).

Question 4. Explain how the test lamp pin is used to test the output of the display unit._____

_____

_____

Step 8. Clear your work area. Properly store equipment and supplies.

Student Activity Sheet 21-1

Review

Name_____ **Score**_____

Date_____ **Class/Period/Instructor**_____

Complete the statements below by filling in the missing word or words.

1. An oscillator must have some form of _____ that is in phase and _____ energy lost by circuit resistance.

   1._____

   _____

2. In an Armstrong oscillator, the _____ _____ determines the frequency of the oscillations.

   2._____

3. The Colpitts oscillator uses _____ _____ for a source of feedback.

   3._____

4. In a Pierce oscillator circuit, the crystal is used in place of the _____ in the tank circuit.

   4._____

5. A crystal's size and thickness will determine its _____.

   5._____

6. The frequency of a crystal will vary in direct proportion to _____.

   6._____

7. A(n) _____ oscillator can convert dc to ac, or ac to dc.

   7._____

8. A push-pull oscillator will have a natural slight _____ that will start the circuit oscillating.

   8._____

9. The Wien bridge op amp oscillator uses _____ and _____ feedback.

   9._____

   _____

10. The abbreviation VCO stands for _____ _____ _____.

    10._____

11. Use a pencil to connect the circuit below as a crystal controlled Hartley oscillator.

Hartley Oscillator

Name_____ **Score**_____

Date_____ **Class/Period/Instructor**_____

Introduction

In this laboratory activity, you will construct a Hartley oscillator. Oscillator circuits are the key component for radio and television transmission, cordless telephones, beepers, cellular phones, satellite transmission, microwave transmission, short wave radio, etc. It is essential to have a basic understanding of several of the basic oscillator circuits. An oscillator electronically converts dc to an alternating current, rather than using a mechanical generator. Oscillators can produce very high frequency ac that is far beyond the range of any conventional mechanical generator. In this lab, you will construct a tapped coil from magnet wire and PVC pipe. Once this is accomplished, you will wire the tapped induction coil into a Hartley oscillator circuit capable of generating over 800 kHz oscillations.

Materials and Equipment

(1)—10 Vdc supply

(1)—110 feet of #30 AWG magnet wire

(1)—several inches of #22 AWG bare wire

(1)—4″ of 3/4 inch PVC pipe

(1)—1 pF capacitor

(2)—47 pF capacitors

(1)—100 pF capacitor

(1)—MPS2222A, NPN transistor

(1)—oscilloscope

(1)—breadboard

Procedure

Step 1. First, prepare the tapped inductive coil. You will need the 4″ of 3/4 PVC pipe and 10 feet of #30 AWG magnet wire. Drill three 1/32 inch holes in the PVC pipe at the locations indicated on the drawing. Secure one end of the magnet wire to the pipe using tape, leaving approximately 4 inches of free conductor. Then, wind 70 turns of the #30 magnet wire around the pipe. You should be at the center hole. Double back approximately 8 inches of the magnet wire, and feed it into the center hole.

Secure the wire at the center hole with a small piece of tape. Continue to wrap the remaining 30 turns of wire around the PVC pipe. Secure the end of the 30 turns.

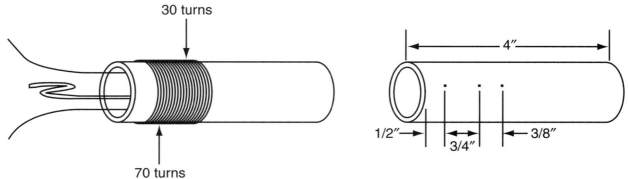

30 turns

70 turns

4″

1/2″   3/4″   3/8″

Step 2. Gently remove the varnish from the outside of the #30 magnet wire ends using a fine grade of sand paper. Remove the varnish until the conductor is bare for 2 inches from each end.

Step 3. Solder the ends to a short piece of #22 bare wire to provide a way to attach the ends of the coils to a breadboard.

Step 4. Now test your coils by taking resistance readings. The 30 turn coil should equal approximately 1.0 ohms, and the 80 turn coil should equal approximately 2.3 ohms for a total of 3.3 ohms. If the resistance is equal to infinity or zero, repair or reconstruct the coils before proceeding further in the lab.

Step 4. Construct the Hartley oscillator circuit that follows. Keep connection wires very short. There will be high frequency oscillations present that can cause a distortion of the output signal. Have your instructor check your circuit.

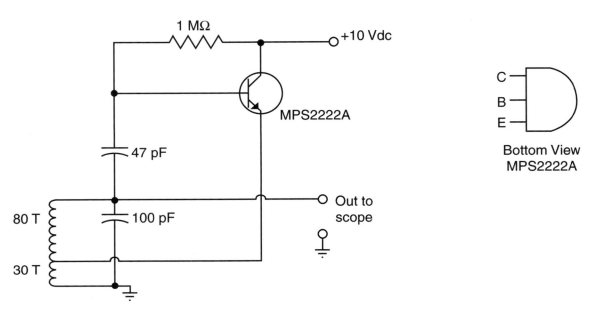

Bottom View
MPS2222A

Step 5. Energize the circuit and connect the oscilloscope to see the output oscillations. The scope sweep should be set to approximately 0.5 microseconds. Connect the ground of the scope close to the oscillator circuit.

Question 1. What is the frequency output of the oscillator? _____

Question 2. Is the output wave pattern a sine wave, square wave, or triangular wave? _____

Question 3. Does this oscillator change dc to ac? _____

Step 6. Replace the 100 pF capacitor connected across the inductor with a 47 pF capacitor, and observe the change on the oscilloscope.

Question 4. How did reducing the capacitance across the inductor affect the output of the oscillator circuit?

_____

_____

Question 5. Which parts of the circuit comprise the tank circuit?_____

_____

_____

Step 7. Clear your work area. Properly store equipment and supplies.

Colpitts Oscillator

Name _____ **Score** _____

Date _____ **Class/Period/Instructor** _____

Introduction

In this laboratory activity, you will construct a typical Colpitts oscillator, and compare it to the Hartley built in the last laboratory. The Colpitts oscillator does not use a tapped coil, but rather two capacitors of unequal value grounded in the center.

Materials and equipment

    (1)—10 Vdc power supply

    (1)—100 μH RF choke

    (1)—MPS2222A NPN transistor

    (1)—10 kΩ resistor, 1/4 W (brown, black, orange)

    (1)—100 kΩ resistor, 1/4 W (brown, black, yellow)

    (1)—1 MΩ resistor, 1/4 W (brown, black, green)

    (1)—0.01 μF capacitor

    (1)—470 pF capacitor

    (1)—100 pF capacitor

    (1)—oscilloscope

    (1)—breadboard

Procedure

Step 1. Gather all materials required for the project. Calibrate the oscilloscope, and set it for a 0.5 μs time sweep.

Step 2. Construct the schematic that follows. Do not apply power until your instructor has checked your work.

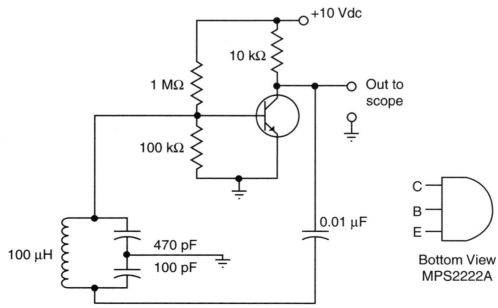

Step 3. After your instructor approves your circuit, turn on the power supply, and observe the output with an oscilloscope.

Question 1. What is the frequency of the sine wave? _____ Hz

Step 4. Slowly lower the supply voltage while observing the output sine wave display on the oscilloscope.

Question 2. How does a varying supply voltage affect the frequency of the oscillator?_____

_____

_____

Question 3. Which parts of the oscillator circuit directly affect the frequency of the output? _____

_____

_____

Question 4. How is feedback for the tank circuit obtained? _____

_____

_____

_____

_____

Question 5. Compare the Colpitts oscillator to the Hartley. What are the similarities? _____

_____

_____

_____

_____

_____

_____

Step 5. Clear your work area. Properly store equipment and supplies.

Wien Bridge Oscillator

Name_____ **Score**_____

Date_____ **Class/Period/Instructor**_____

Introduction

In this lab activity, you will construct and experiment with a Wien bridge oscillator. The Wien bridge oscillator is unique in the fact that it uses a simple incandescent lamp as part of its feedback circuit to help maintain a steady waveform. Look closely at the schematic of the Wien bridge oscillator. Note that the feedback circuit consists of a lamp, a fixed resistor, and a variable resistor. The variable resistor will be used to calibrate the amount of feedback in the circuit, and the lamp will maintain it. As the lamp conducts, it will warm up, increasing the lamp filament resistance. The lamp resistance increases the current. As the current through the lamp reduces, so will the resistance of the lamp filament. The lamp maintains an electrical balance, keeping the op amp gain locked into step. The circuit will not work without the lamp. Frequency of the output is controlled by the two matching resistors and two matching capacitors. One pair is connected in series, and the other in parallel. By varying the value of the resistors or capacitors, you can vary the frequency of the output.

Materials and Equipment

(1)—741 op amp

(1)—14-volt lamp

(1)—lamp socket

(3)—10 kΩ resistors, 1/4 W (brown, black, orange)

(2)—1 kΩ resistors, 1/4 W (brown, black, red)

(2)—100 kΩ resistors, 1/4 W (brown, black, yellow)

(1)—50 kΩ potentiometer

(2)—0.022 μF capacitors

(1)—oscilloscope

(2)—9-volt batteries and holders

(1)—breadboard

Procedure

Step 1. Gather all materials and equipment necessary for the laboratory. Calibrate your oscilloscope.

Step 2. Assemble the schematic that follows. However, do not energize the circuit until it is inspected by your instructor. Resistors R_1 and R_2 should be 1 kΩ. Capacitors C_1 and C_2 should be 0.022 μF.

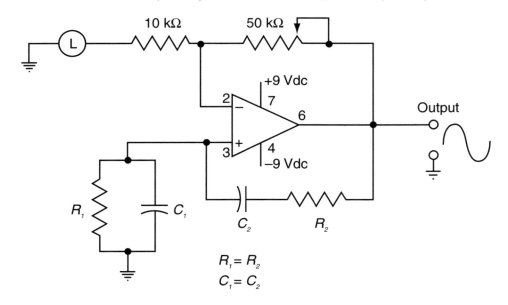

Step 3. Energize the circuit with the 50 kΩ potentiometer set at approximately half value. The oscilloscope time sweep should be at or around 50 μs. The volt/div. should be at or around 0.5 using a ×10 probe.

Step 4. Slowly adjust the 50 kΩ potentiometer until a waveshape appears on the scope. The waveshape will probably resemble a square wave. This is because the sine wave is being clipped. Slowly reduce the 50 kΩ potentiometer until a perfect sine wave appears. If you adjust the potentiometer too far, the wave will disappear and become flat. Slowly raise the value once more.

Step 5. What is the frequency of the sine wave with a 1 kΩ resistor and 0.022 μF capacitor in the bridge of the circuit?

Record value here for 0.022 μF and 1 kΩ = _____ Hz

Step 6. Turn off the power, and change the value of resistors R_1 and R_2 from 1 kΩ to 10 kΩ. Record the frequency. 0.022 μF and 10 kΩ = _____ Hz

Step 7. Turn off the power, and change the value of resistors R_1 and R_2 to 100 kΩ.

Record the frequency. 0.022 μF and 100 kΩ = _____ Hz

Question 1. What effect does the resistance value have on the output frequency? _____

_____

_____

Question 2. What effect would changing the size of capacitors C_1 and C_2 have on the output frequency?

_____

_____

Step 8. Clear your work area. Properly store equipment and supplies.

AM and FM Radio Communications

Student Activity Sheet 22-1

Review

Name _____ **Score** _____

Date _____ **Class/Period/Instructor** _____

Complete the following sentences by filling in the missing word or words.

1. In the simple radio receiver, the _____ circuit is the circuit that selects the radio station.

1. _____

2. A radio wave consists of _____ radiation.

2. _____

3. The electromagnetic wave is transmitted perpendicular to the _____ wave.

3. _____

4. Radio waves are divided into two groups, _____ waves and _____ waves.

4. _____
   _____

5. Radio waves travel through space at _____ miles per hour or _____ meters per second.

5. _____
   _____

6. The symbol λ stands for _____.

6. _____

7. The wavelength of a 93 kHz radio wave is _____ miles.

7. _____

8. The wavelength of a 4.5 MHz radio wave is _____ feet.

8. _____

9. The letters FCC stand for _____ _____ _____.

9. _____

10. AM radio uses the _____ frequency range while television uses the _____ and _____ frequency ranges.

10. _____
    _____
    _____

11. A(n) _____ connected to an antenna will create radio waves.

11. _____

12. The oscillator creates the signal known as the _____ wave.

12. _____

13. The letters CW stand for _____ _____.

13. _____

14. A sound wave can be converted into a(n) _____ _____ wave by utilizing a microphone.

14. _____

15. Combining the carrier wave with an audio wave is called _____.

15. _____

(Continued)

16. The letters AM stand for _____ _____ and the letters FM stand for _____ _____.

16. _____
_____

17. Combining a 5 kHz and a 2000 kHz wave will produce a(n) _____ kHz wave and a(n) _____ kHz wave.

17. _____
_____

18. The two frequency waves added together are called the _____ sideband while the two frequencies that are subtracted form the _____ sideband.

18. _____
_____

19. The AM band extends from _____ kHz to _____ kHz.

19. _____
_____

20. Each AM broadcast band is _____ kHz.

20. _____

21. The FM broadcast band is from _____ MHz to _____ MHz.

21. _____
_____

22. The amount of variation away from the center frequency of an FM band is called _____ _____.

22. _____

23. The audio _____ _____ frequency determines the rate of frequency deviation.

23. _____

24. The maximum _____ _____ and the maximum _____ _____ determine the modulation index.

24. _____
_____

25. The removal of the audio signal from the carrier wave is called _____ or _____.

25. _____
_____

26. The letters TRF stand for _____ _____ _____.

26. _____

27. The ability of a receiver to select a single frequency is called _____ and the ability of a radio receiver to pick up a weak signal is called _____.

27. _____
_____

28. Heterodyning means _____ of _____.

28. _____
_____

29. The superheterodyne receiver uses a beat frequency equal to _____ kHz.

29. _____

30. When two signals are mixed together, the result will be four signals. If a 500 kHz signal is mixed together with a 800 kHz signal, the value of the four resulting signals will be _____ kHz, _____ kHz, _____ kHz, and _____ kHz.

30. _____
_____
_____
_____

31. The _____ converts the incoming radio signal to a intermediate frequency.

31. _____

32. In AM broadcasting, _____ kHz is the best intermediate frequency and for FM it is _____ MHz.

32. _____

_____

33. The letters AGC stand for _____ _____ _____.

33. _____

34. The limiter is designed to keep the radio wave _____ from varying.

34. _____

35. Tone controls can vary the audio tone by adjusting the amount of high or low _____ in the speaker circuit.

35. _____

Crystal Radio Receiver

Name_____ **Score**_____

Date_____ **Class/Period/Instructor**_____

Introduction

In this laboratory activity, you will construct a simple AM radio receiver. This radio receiver is known as a crystal radio. Take note of the fact that it uses no battery or any other type of power supply. The power used for this radio is supplied by the electrostatic field transmitted by the radio station. Two things are critical for the radio to work. One is a reliable antenna, and the other is a good ground. Without the antenna and ground, the system will never work properly. You can use a 40 foot length of #22 conductor for an antenna. The ground can be the ground at your power supply. If the project antenna is constructed outside your laboratory, a two to three foot length of 1/2″ pipe or metal rod driven into the earth will make a good ground for the system.

Examine the schematic that follows. C_1 is a 360 pF variable capacitor. When the variable capacitor is coupled together with the 100 turn tapped coil you made, they form a tank circuit. The tank circuit forms the tuner for the AM radio receiver. The diode is used to eliminate 1/2 of the transmitted wave or rectify the signal. Without the diode, the earphone speaker could not reproduce the transmitted sounds. The C_2 is a 47 pF capacitor used to help filter the signal to the speaker. The 100 kΩ resistor is used to help load the diode which also assists in the filtering action.

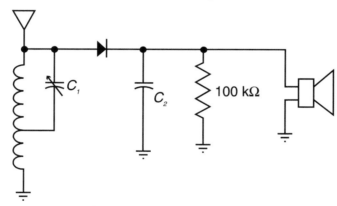

Materials and Equipment

  (1)—germanium diode 1N34A

  (1)—earphone

  (1)—variable capacitor 360 pF

  (1)—47 pF capacitor, 50 V

  (1)—100 kΩ resistor, 1/4 W (brown, black, yellow)

  (1)—antenna (at least 40 feet of #22 conductor)

  (1)—good grounding source (2 feet of rod or pipe)

  (1)—oscilloscope

  (1)—breadboard

Procedure

  Step 1. Gather all materials needed for the laboratory activity. Calibrate your oscilloscope. The 40 feet of antenna wire may not be needed. An antenna system and/or grounding source may be provided by your instructor.

Chapter 22   AM and FM Radio Communication   **279**

Step 2. Assemble the crystal radio receiver according to the schematic.

Step 3. Place the earphone to your ear to detect any radio stations. The signal may be faint and require silence in the surrounding area. Slowly rotate the variable capacitor to tune the tank circuit to detect and enhance any radio stations.

Step 4. Connect the oscilloscope to the circuit between the diode and the tank circuit. Observe the sine wave pattern. The oscilloscope volts/div. will need to be set very low (approximately 0.005) and the time/div. to approximately 0.5 milliseconds.

Step 5. Try different configurations of the antenna, tuning capacitor, and ground in the tank circuit. Note any improvement in the signal.

Step 6. Clear your work area. Properly store equipment and supplies.

Crystal Radio Receiver and LM386 Audio Amplifier

Name_____ **Score**_____

Date_____ **Class/Period/Instructor**_____

Introduction

In this laboratory activity, you will improve the crystal radio receiver from the last project through the addition of a low power audio amplifier chip. The signal received by the AM radio you built in the last lab activity will be amplified approximately 200 times by the addition of a LM386 audio amplifier chip.

The 10 kΩ potentiometer connected to pin 3 acts as a volume control. The signal from the radio receiver is input to the 10 kΩ potentiometer. The output to the earphone is connected to pin 5 through a 33 μF capacitor. Use caution when placing the earphone into your ear; the output can be quite loud. The amount of amplification is controlled by the capacitor across pins 1 and 8. This configuration will produce a gain of approximately 200. Without the capacitor across pins 1 and 8, the gain would be approximately 20. The 100 μF capacitor connected to pin 7 is a bypass capacitor and helps to clear the signal. Power is applied to pin 6 and can vary up to 15 volts maximum. The maximum output power for the LM386 is 400 milliwatts. Exceeding these values will destroy the amplifier.

Materials and Equipment

(1)—0-12 Vdc power supply

(1)—germanium diode 1N34A

(1)—LM386 audio amplifier

(1)—variable capacitor 360 pF

(1)—10 kΩ potentiometer

(1)—100 kΩ resistor, 1/4 W (brown, black, yellow)

(1)—47 pF capacitor, 50 V

(1)—10 μF capacitor, 50 V

(1)—33 μF capacitor, 50 V

(1)—100 μF capacitor, 50 V

(1)—earphone

(1)—antenna, (at least 40 feet of #22 conductor)

(1)—good grounding source (2 feet of rod or pipe)

(1)—oscilloscope

(1)—breadboard

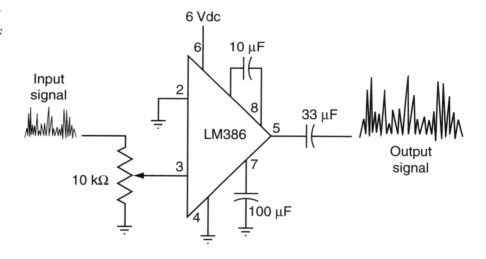

Procedure

Step 1. Assemble all materials required for the project. Calibrate the oscilloscope.

Step 2. Construct the circuit in the schematic that follows. The input to the LM386 is from the output of the crystal radio receiver you constructed in Student Activity Sheet 22-2. Connect the input to the same location the earphone connected to in the radio receiver. The earphone will connect to the output of the audio amplifier.

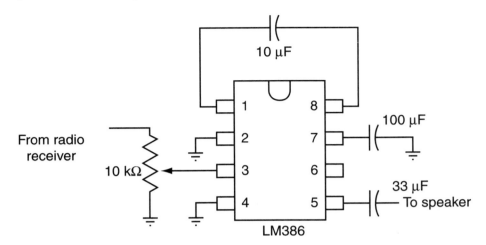

Step 3. Rotate the capacitor until the strongest and clearest radio station is heard. Caution should be used. The sound produced in the earphone can be very loud depending upon the proximity of the radio station broadcasting the signal.

Step 4. Connect the oscilloscope to the input of the audio amplifier, and then connect it to the output of the audio amplifier. Compare the signals. Observe the output signal while listening to the sounds and determine if they are in step with each other.

Question 1. Does the oscilloscope display signal amplitude fluctuate in step with the sounds? _____

Step 5. Place a jumper across the diode. Listen through the earphone.

Question 2. Did the radio receiver still work? _____

Question 3. If the radio receiver still worked, what caused the rectification of the signal? _____

_____

Step 6. Clear your work area. Properly store equipment and supplies.

Transmitting Sound Waves through Electrical Conductors

Name_____ **Score**_____

Date_____ **Class/Period/Instructor**_____

Introduction

In this lab activity, you will build a communication system based on the transmission of an audio signal through a conductor. Many electronic systems, such as the telephone, PA systems, intercoms, musical sound systems, etc., operate on the principle of sound transmitted by a conductor. The circuit consists of a condenser microphone which will change your voice vibrations into electrical variations. The electrical variations leaving the microphone will first be amplified by a 741 op amp and then through a 386 audio amplifier. Once the signal has been sufficiently amplified by the op amp and the 386 audio amplifier, the signal is sent through a copper wire to the 8 Ω speaker. The speaker will vibrate in sequence with the frequency and amplitude of the audio signal. The vibration in the speaker will vibrate the air surrounding the speaker in step with the audio pattern generated at the microphone. The vibrating air strikes the inner ear and is interpreted as sound by the brain.

Materials and Equipment

(1)—0-12 Vdc power supply

(1)—dual voltage power supply 9 Vdc (or two 9-volt batteries)

(1)—741 op amp

(1)—condenser microphone

(1)—LM386 audio amplifier

(1)—8-ohm speaker

(2)—220 Ω resistors, 1/4 W (red, red, brown)

(2)—1 kΩ resistors, 1/4 W (brown, black. red)

(1)—0.2 kΩ resistor, 1/4 W (brown, red, red)

(1)—10 kΩ potentiometer

(1)—100 kΩ potentiometer

(1)—1 µF capacitor, 50 V

(2)—10 µF capacitors, 50 V

(2)—100 µF capacitors, 50 V

(1)—oscilloscope

(1)—breadboard

Bottom view of condenser microphone

Side view

Procedure

Step 1. Assemble all required materials, and calibrate the oscilloscope.

Step 2. Construct the simple circuit in the schematic that follows. Take special note of the location of the ground pin on the condenser microphone. The ground pin connects to the outside casing of the microphone. The positive voltage connects to the remaining pin. Do not reverse these connections in the circuit. Have your instructor inspect your circuit before you energize it.

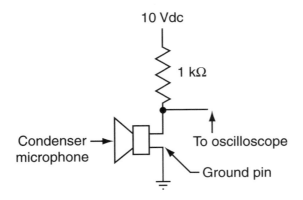

Step 3. Connect the oscilloscope to the output of the microphone. The scope should be set at or about 0.02 µs time sweep and 0.02 volts/div. Turn on the power supply, and set it for 10 Vdc.

Step 4. With the oscilloscope connected to the output of the microphone, gently rub your finger tip across the microphone surface. This should produce variations in a sine wave pattern that are observable on the scope. If not, call your instructor.

Step 5. Disconnect the power, and connect the amplifier and the speaker to the microphone as indicated in the schematic that follows. Set the 10 kΩ potentiometer to approximately 1/2 total resistance, or midway rotation. Have your instructor check it before energizing the circuit.

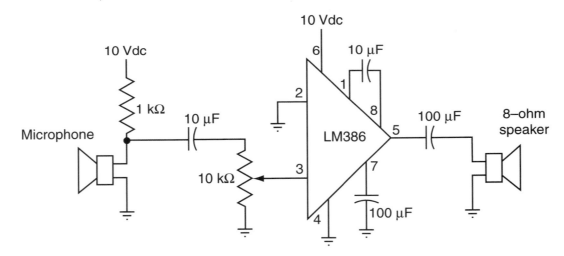

Step 6. Energize the circuit, and gently rub the surface of the microphone with your finger tip. Connect the oscilloscope to the output, and observe the wave pattern displayed on the scope.

Step 7. Adjust the 10 kΩ potentiometer downward if there is too much hum or feedback from the speaker.

Step 8. Gently blow into or across the microphone, and listen for the sound produced at the 8-ohm speaker. If there is no sound produced, check your connections closely before calling your instructor for help.

Electronic Communication and Data Systems

Step 9. Add the 741 op amp to the circuit using the following schematic. There have been some modifications to the circuit to enhance the performance. The op amp has a 9-volt positive and a 9-volt negative supply required for the circuit. The microphone has two 220-ohm resistors added as a voltage divider circuit to supply 4.5 volts to the microphone. The microphone operates at its optimum when connected to 4.5 volts rather than 9 volts. Also, do not connect the 10 μF capacitor between pins 1 and 8 on the 386 audio amp. This will be done later in the lab.

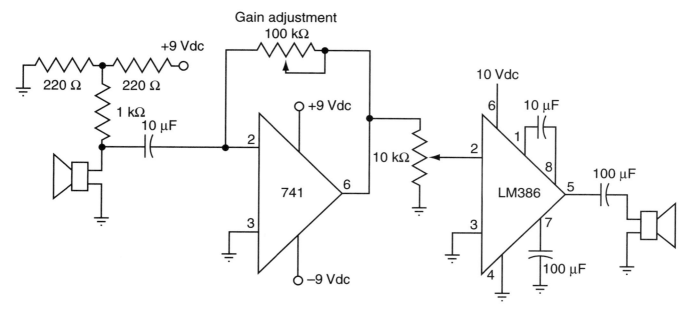

Step 10. Start both potentiometers at mid-range, and slowly proceed to adjust each until optimum performance is heard. Try counting into the microphone while adjusting the potentiometers. The presence of a load hum or popping sound can be relieved by adjusting the op amp gain potentiometer. If you encounter a lot of problems adjusting the potentiometers, try connecting a signal generator to the input of the op amp. Set the frequency to approximately 1 kHz and the voltage as low as possible. This will simulate the voltage produced by the microphone.

Step 11. After you have achieved a quality sound from the speaker, raise the output gain of the 386 audio amplifier by inserting a 1.2 kΩ resistor in series with a 10 μF capacitor between pin 1 and pin 8 of the 386 audio amp. This should increase the gain to 50. Without the addition of the capacitor and the 1.2 kΩ resistor the nominal gain is only 20. You will need to readjust the potentiometers each time you modify the circuit.

Step 12. Now you will raise the gain of the audio amplifier to 200 by removing the 1.2 kΩ resistor and leaving only the 10 μF capacitor between pins 1 and 8 on the 386. You will need to readjust the potentiometers once again after this modification.

Question 1. Design a circuit that will simulate an intercom or two-way communication. You may use an additional microphone, speaker, switches, and wire. There is no need for additional amplifiers. Draw the schematic in the following space.

Step 13. Clear your work area. Properly store equipment and supplies.

AM Radio Transmitter

Name _____ **Score** _____

Date _____ **Class/Period/Instructor** _____

Introduction

In this laboratory activity, you will construct a simple AM transmitter. To transmit an audio signal through the air, you need to construct an oscillator that operates in the AM radio receiver band. The oscillator will produce the carrier wave needed for transmission of the audio sound. The audio sound is produced by a condenser microphone and simple, one-transistor amplifier circuit. The audio circuit is tied into the oscillator circuit and will cause the carrier wave signal to be distorted by the audio signal. The receiver will separate the audio signal from the carrier wave and reproduce the sound through the receiver speaker system.

Materials and Equipment

(1)—6-volt dc power supply

(1)—100 µH RF choke

(1)—condenser microphone

(2)—MPS222A NPN transistors

(2)—1 kΩ resistors, 1/4 W (brown, black, red)

(1)—10 kΩ resistor, 1/4 W (brown, black, orange)

(1)—100 kΩ resistor, 1/4 W (brown, black, yellow)

(2)—1 MΩ resistors, 1/4 W (brown, black, green)

(2)—0.01 µF capacitors, 50 V

(1)—1 µF capacitor, 50 V

(1)—2.2 µF capacitor, 50 V

(1)—100 pF capacitor, 50 V

(1)—470 pF capacitor, 50 V

(1)—portable AM/FM radio

(1)—oscilloscope

(1)—breadboard

Procedure

Step 1. Gather all required materials and equipment. Calibrate your oscilloscope.

Step 2. Construct the circuit that follows. Have your instructor check the circuit for accuracy before you energize. Use approximately 18" of conductor as the antenna.

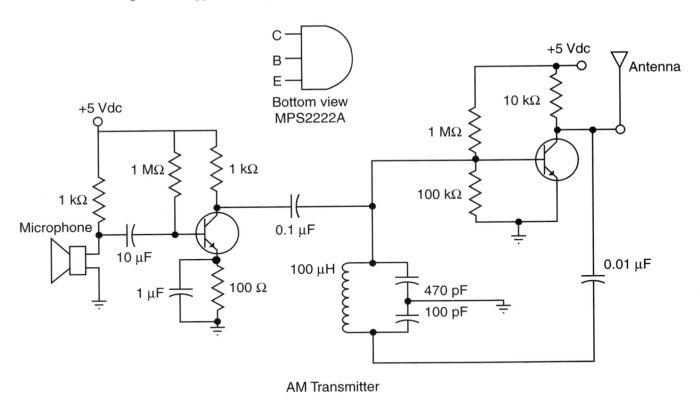

AM Transmitter

Step 3. Turn on the power, and adjust the volume of the portable radio to approximately half value. Slowly tune the radio, and listen for a low steady hum. When you think you have located the hum produced by the transmitter, turn the power off to the transmitter and see if the hum stops. Turn the power back on, and the hum should return. If the hum is directly over a broadcasting radio station, change the size of the 470 pF and 100 pF capacitors to higher or lower values. By changing the value of the capacitors, the transmitter oscillator will produce a different frequency. This results in a different station assignment.

Step 4. Speak into the microphone and listen to the speaker. You may need to have someone else listen to the speaker. Vary the tuner over the radio receiver to find optimum performance.

Step 5. Connect the oscilloscope across the oscillator circuit. Observe the wave pattern shift when you speak into the microphone.

Step 6. Determine the maximum range of the AM transmitter you have built.

Step 7. If a second AM radio is available, you can use the two AM radio receivers to locate the position of the transmitter. Rotate the radio receiver in a stationary circle. Note where the received signal seems stronger in a certain direction. The strength of the signal is directly related to the position of the antenna of the receiver. You can use two radios to locate the position of the transmitter. Try it now.

Question 1. A homing device produces a repeating beep sound on a radio receiver. How can a homing device be built from this project of any radio transmitter?

_____

_____

_____

Electronic Communication and Data Systems

Question 2. Where would you install a variable capacitor to readily change the frequency of the transmitter?

_____

_____

_____

Question 3. How could the circuit be modified to produce a stronger output signal? _____

_____

_____

_____

Step 8. Clear your work area. Properly store equipment and supplies.

FM Radio Transmitter

Name _____ **Score** _____

Date _____ **Class/Period/Instructor** _____

Introduction

In this laboratory activity, you will construct and test an FM transmitter. The circuit of the FM transmitter is similar to the AM transmitter built in the Student Activity Sheet 22-5. The major difference is in the tank circuit of the oscillator used to produce the carrier wave. FM is a much higher frequency than AM, so the tank circuit must be able to produce a much higher frequency. The FM transmitter will produce a better quality sound than the AM transmitter in the last circuit.

Materials and Equipment

(1)—3 Vdc power supply

(2)—MPS2222A transistors

(1)—18″ length of #14 AWG bare solid conductor (used to make coil)

(1)—condenser microphone

(1)—100 Ω resistor, 1/4 W (brown, black, brown)

(1)—120 Ω resistor, 1/4 W (brown, red, brown)

(2)—1 kΩ resistors, 1/4 W (brown, black, red)

(1)—10 kΩ resistor, 1/4 W (brown, black, orange)

(2)—47 kΩ resistors, 1/4 W (yellow, violet, orange)

(2)—1 MΩ resistors, 1/4 W (brown, black, green)

(1)—0.1 μF capacitor, 50 V

(1)—1 μF capacitor, 50 V

(1)—10 μF capacitor, 50 V

(1)—4 pF capacitor, 50 V

(1)—47 pF capacitor, 50 V

(1)—oscilloscope

(1)—portable AM/FM radio

(1)—breadboard

Procedure

Step 1. Gather all materials needed for the laboratory activity.

Step 2. You will now construct the coil used for the oscillator circuit in the FM transmitter. Carefully study the illustration that follows before you begin. The internal diameter of the coil should be between 1/2 and 5/8 of an inch. The coil could be wound over a dowel rod of the same diameter or any tubular shaped object such as a permanent marker. It can even be paper rolled around a pencil until the desired diameter is reached. There should be six turns or loops in the total coil. A tap will be placed at the second coil and should be equal to one and one half to two full turns. The overall length of the complete coil should not exceed 3/4 of an inch. Three #22 AWG leads need to be soldered to the coil for easy attachment to the breadboard—one lead at each end of the coil, and the third is the tap one and one half to two turns from one end of the coil. These specifications should be closely followed.

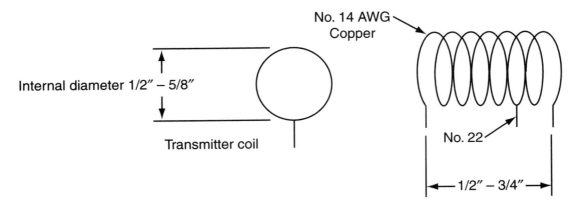

Step 3. Have your instructor inspect the coil before proceeding further in this project.

Step 4. Once your coil is complete, proceed to construct the schematic that follows.

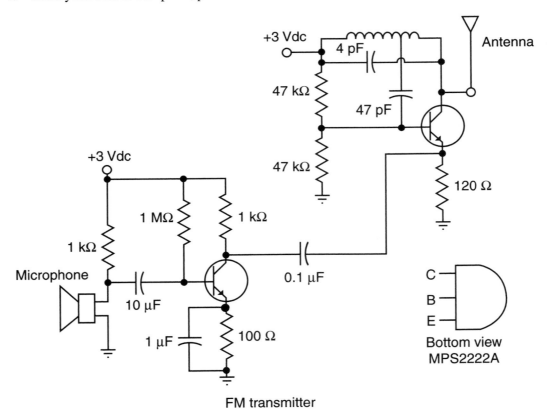

FM transmitter

Step 5. Energize the circuit with 3 Vdc.

Step 6. Turn on a FM radio, and slowly adjust the station dial until a steady low hum is heard. The steady low hum will identify the transmitter frequency in relation to the station identified on the FM scale. To verify the match of the frequencies, turn the transmitter off and then back on. If the transmitter is aligned with an existing station, change the value of the 47 pF capacitor in the oscillator circuit to a higher value. This will change the transmission frequency.

Step 7. Speak into the microphone, and listen to the FM radio. Slowly adjust the radio tuner until optimum performance of the transmitter is found.

Step 8. If a second radio receiver is available, use triangulation to locate the FM transmitter.

Question 1. What is the major difference between the AM transmitter in Student Activity Sheet 22-5 and the FM transmitter? _____

_____

_____

_____

Question 2. How could the strength of the transmitter be increased? _____

_____

_____

_____

Step 9. Clear your work area. Properly store equipment and supplies.

Student Activity Sheet 23-1

Review

Name_____ **Score**_____

Date_____ **Class/Period/Instructor**_____

Complete the following sentences by filling in the missing word or words.

1. The scanning system used in the US for commercial television consists of _____ lines.

    1. _____

2. The _____ numbered lines are scanned first and then the _____ numbered lines.

    2. _____

    _____

3. A field consists of one scan of _____ lines.

    3. _____

4. A TV screen displays _____ frames per second, or _____ fields per second.

    4. _____

    _____

5. The coils in a picture tube that sweep the electron beam across the inside of the tube are called the _____ _____.

    5. _____

6. The horizontal oscillator operates at _____ Hz.

    6. _____

7. A(n) _____ _____ is transmitted along with the picture information to keep the scanning on the TV screen in step with the camera at the studio.

    7. _____

8. _____ amplitudes create dark spots of the screen image while _____ amplitudes create white or lighter spots of the screen image.

    8. _____

    _____

9. The invention that made color TV possible was the _____ _____ picture tube.

    9. _____

10. A(n) _____-_____ tube uses three electron guns for color TV.

    10. _____

11. Each TV channel is _____ MHz wide.

    11. _____

12. The letters UHF stand for _____ _____ _____ while the letters VHF stand for _____ _____ _____.

    12. _____

    _____

13. VHF channels are from _____ to _____ while UHF channels are from _____ to _____.

    13. _____

    _____

(Continued)

14. In addition to video and audio information, a VHF tape also contains _____ information. (List four items.)

14. _____

_____

_____

_____

15. _____ _____ _____ first introduced the idea of using satellites for communications.

15. _____

16. If a satellite stayed directly over a city even while the earth is rotating, the satellite would be in a(n) _____ orbit.

16. _____

17. The signals from a satellite are concentrated at the _____ point on a satellite dish receiver.

17. _____

18. Another name for the satellite dish is _____ antenna.

18. _____

TV Remote Control

Name_____ **Score**_____

Date_____ **Class/Period/Instructor**_____

Introduction

In this laboratory activity, you will experiment with a typical infrared remote control receiver. The main device of the circuit is a GP1U52X infrared receiver/demodulator. The receiver is a self-contained unit consisting of several individual components combined to form a single unit. The first component is a photodiode that detects the infrared beam from a transmitter. The infrared beam is not within the spectrum of light that can be detected by the human eye. The electrical pulse from the photodiode detector passes through many stages. These include an amplifier, band-pass filter, and demodulator. These components are necessary to reconstruct and shape the electrical energy into a near perfect representation of the digitized infrared beam that was received by the GP1U52X detector unit. In this laboratory you will build a simple circuit that will allow you to see the digitized signal displayed on an oscilloscope.

Materials and Equipment

    (1)—5 Vdc power supply

    (1)—GP1U52X infrared
        receiver/demodulator

    (1)—remote control transmitter
        (typical TV remote)

    (1)—1 MΩ resistor, 1/4 W
        (brown, black, green)

    (1)—oscilloscope

    (1)—breadboard

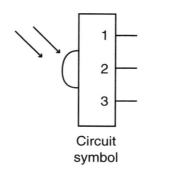 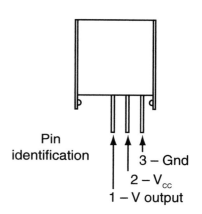

Circuit symbol   Pin identification   3 – Gnd

2 – V_{cc}

1 – V output

GP1U52X infrared detector/demodulator

Procedure

    Step 1. Gather all materials, and calibrate the oscilloscope.

    Step 2. Construct the simple circuit that follows according to the schematic provided. Have your instructor check the circuit before energizing.

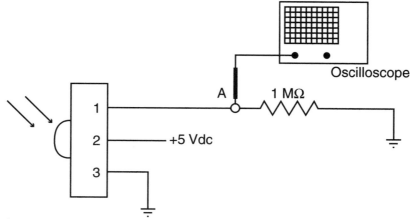

Chapter 23   Television

Step 3. Energize the circuit with 5 Vdc. Set the oscilloscope adjustments on or about 0.2 volts/div. Set the time/div. on or about 0.5 ms.

Step 4. Aim the infrared transmitter toward the infrared detector unit. A pattern should be displayed on the oscilloscope screen. The pattern should resemble a series of dots and dashes. If it does not, adjust the oscilloscope until such a pattern is displayed.

Step 5. Next, lower the time/div. range in small increments until a square wave pattern fills the screen. Continue to vary the time/div. between the displayed pattern of dots and dashes until it is clear to you that the string of dots and dashes displayed on the oscilloscope is actually a string of square wave pulses.

Question 1. When the infrared beam receiver is activated by the transmitted infrared beam, does the output of the detector go high or low? _____

Question 2. Press several buttons on the infrared transmitter and closely observe the pattern of dots and dashes displayed. How do the patterns displayed vary when different buttons are pressed on the remote transmitter?

_____

_____

Step 6. Design a circuit that will cause a lamp or LED to light or go dark when the infrared detector is activated by the use of a remote control infrared transmitter. Draw the block diagram below.

Step 7. Clear your work area. Properly store equipment and supplies.

Calculate and Construct a TV Antenna

Name _____ **Score** _____

Date _____ **Class/Period/Instructor** _____

Introduction

In this laboratory activity, you will design a television antenna to receive local stations. It is a simple activity that will reinforce the concept of wavelength. You will calculate the length of the antenna receiver necessary for local station reception. The wavelength of a transmitted signal is based on frequency of the transmitted signal and the speed of the electromagnetic wave. The formula for calculating wavelength is below.

$$\text{wavelength } (\lambda) = \frac{984 \text{ feet per second (or 300 meters per second)}}{\text{frequency (MHz)}}$$

To calculate the length of antenna needed for a 20 MHz transmitter, simply divide 984 by 20 to find the length in feet or divide 300 by 20 to get the length in meters. The answer is 49.2 feet or 15 meters. The actual design should be approximately 5% shorter than the theoretical length. By shortening the length 5% you will reduce the amount of self-induction in the antenna system.

Your project will be to design a simple dipole antenna system to be used on a television. You will identify the nearest television stations in your area and determine the frequency of those stations using the chart provided. You will be cutting lengths of metal rod, tubing, or heavy gauge conductor to represent the length of antenna elements needed. You will then connect the transmission lead-in wire to the antenna elements and then to the television. Next, the antenna will be positioned where the strongest reception is possible. Finally, you will compare the reception of the television between the station for which the antenna was designed and another station that would require a longer or shorter antenna element.

Materials and Equipment

(1)—television

(1)—two-wire ribbon flat lead

(1)—length of heavy conductor or 1/2″ or smaller tubing

(1)—antenna support

(1)—75 Ω coaxial converter

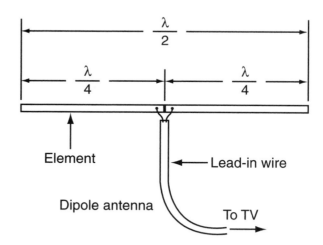

Procedure

Step 1. Select the nearest television station, and identify the frequency of the station transmission. See the chart at the end of this laboratory and match the channel to the midrange frequency. Calculate the transmission length of the electrostatic wavelength using the formula provided in the introduction, and then divide the length of the wave by 4 to determine the element length needed. See the illustration to the right of the materials and equipment.

Length of one element determined by calculation = _____

Length of one element determined by calculation less 5% = _____

Step 2. Mount the two elements so that they are close together in the center but do not touch each other.

Step 3. Connect the lead-in wire to the two elements.

Step 4. Connect the lead-in wire to a television receiver located at the back of the television. You will probably need a 75 Ω converter to connect the twin lead cable to the back of the television since most televisions are made cable ready.

Step 5. Determine the direction that the television signal is strongest. It will most likely be in the direction of the television station the element was designed to receive.

Step 6. Change the channel on the television and see how the reception for the other stations are affected by the antenna you designed.

Step 7. Repeat the experiment for another television station in your area.

Step 8. Clear your work area. Properly store equipment and supplies.

Television Channel Frequencies

CH	MHz	CH	MHz	CH	MHz	CH	MHz	CH	MHz
2	54-60	21	512-518	40	626-632	59	740-746	78	854-860
3	60-66	22	518-524	41	632-638	60	746-752	79	860-866
4	66-72	23	524-530	42	638-644	61	752-758	80	866-872
5	72-76	24	530-536	43	644-650	62	758-764	81	872-878
6	76-82	25	536-542	44	650-656	63	764-770	82	878-884
7	176-180	26	542-548	45	656-662	64	770-776	83	884-890
8	180-186	27	548-554	46	662-668	65	776-782		
9	186-192	28	554-560	47	668-674	66	782-788		
10	192-198	29	560-566	48	674-680	67	788-794		
11	198-204	30	566-572	49	680-686	68	794-800		
12	204-210	31	572-578	50	686-692	69	800-806		
13	210-216	32	578-584	51	692-698	70	806-812		
14	470-476	33	584-590	52	698-704	71	812-818		
15	476-482	34	590-596	53	704-710	72	818-824		
16	482-488	35	596-602	54	710-716	73	824-830		
17	488-494	36	602-608	55	716-722	74	830-836		
18	494-500	37	608-614	56	722-728	75	836-842		
19	500-506	38	614-620	57	728-734	76	842-848		
20	506-512	39	620-626	58	734-740	77	848-854		

Student Activity Sheet 24-1
Review

Name_____ **Score**_____

Date_____ **Class/Period/Instructor**_____

Complete the following sentences by filling in the missing word or words.

1. Fiber-optic systems transmit information through cables in the form of _____.

   1. _____

2. The reasons for using fiber-optic cables are _____. (List six reasons.)

   2. _____

   _____

   _____

   _____

   _____

   _____

3. A fiber-optic cable is constructed of a(n) _____ or a(n) _____ core surrounded by _____, which is surrounded by a buffer area.

   3. _____

   _____

   _____

4. Loss of fiber-optic signal is called _____ and is measured in _____.

   4. _____

   _____

5. Signal loss due to impurities in the fiber-optic cable is called _____.

   5. _____

6. Signal loss associated with the light beam reflecting off the cladding causing the signal to arrive at slightly different times than transmitted is called _____.

   6. _____

7. Bending, splicing, and connectors are classified _____ cable losses.

   7. _____

8. Joining two fiber-optic cable cores together using heat is called _____ _____.

   8. _____

9. When two different materials are spliced together a(n) _____ _____ loss will occur.

   9. _____

10. The electrical signal is changed into a light signal by using _____ or _____ _____.

10. _____

_____

11. A forward biased diode will emit _____ at its junction.

11. _____

12. The letters PIN stand for _____ _____ _____.

12. _____

13. An avalanche photodiode is used as a(n) _____ in a fiber-optic system.

13. _____

14. The letters OTDR stand for _____ _____ _____ _____.

14. _____

15. An OTDR tests fiber-optic cables by transmitting _____ pulses through the cable and then _____ them back to the receiver.

15. _____

_____

16. The display analysis on the screen of a OTDR is called a(n) _____ trace.

16. _____

17. Stimulation of a laser is provided by a(n) _____.

17. _____

18. When the light reflecting inside the laser tube releases additional photons, _____ has occurred.

18. _____

19. _____ protection must be worn when experimenting with lasers.

19. _____

Exploring Infrared Light

Name_____ **Score**_____

Date_____ **Class/Period/Instructor**_____

Introduction

In this laboratory activity, you will explore infrared light and compare it to other sources of light. Because infrared light is invisible to the naked eye, it is impossible to distinguish between an infrared emitter that is working from one that is not by the naked eye alone. In the first part of this laboratory activity you will learn to distinguish between a working and a nonworking infrared LED and infrared detector.

One of the difficulties with detector infrared transistors is the various ways manufacturers identify the anode, cathode, and base (when a base exists). Not all detector infrared transistors have a base lead. Some simply incorporate the base into the transistor and leave only the emitter and collector or anode and cathode. Typical designs of identification are illustrated to the right of the materials and equipment list and should cover most cases.

Materials and Equipment

(1)—0–9 Vdc power supply

(1)—555 timer

(1)—3″ × 5″ card

(1)—pair infrared emitter and detector

(1)—510 Ω resistor, 1/4 W (green, brown, brown)

(1)—47 kΩ resistor, 1/4 W (yellow, violet, orange)

(1)—2.2 kΩ resistor, 1/4 W (red, red, red)

(1)—100 kΩ potentiometer

(1)—1 μF capacitor, 50 V

(1)—0.01 μF capacitor, 50 V

(1)—red LED

(1)—yellow LED

(1)—green LED

(1)—2″ piece of drinking straw

(1)—2″ × 2″ square of aluminum foil

(1)—12-volt incandescent lamp and lamp holder

(1)—voltmeter

(1)—dual trace oscilloscope

(1)—breadboard

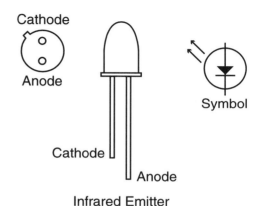

Infrared Emitter

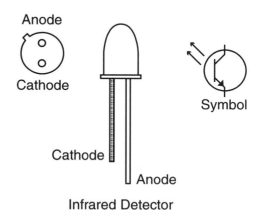

Infrared Detector

Procedure

Step 1. Gather all materials needed for the lab activity. Calibrate the oscilloscope—both channels.

Step 2. Assemble your circuit using the schematic that follows. First bend the leads on the emitter and detector using needle nose pliers. Insert them into the breadboard approximately one and one half inches apart. The tops of the emitter and detector should face each other. The light emitted from the LED is concentrated at the top of the lens. The detector is also most sensitive to light at the top of the lens. Do not bend the leads adjacent to the lens on top of the LED. Bending the leads near the top will cause the leads to weaken and break off. Only bend the leads at the midpoint of their length. Have your instructor inspect your circuit before you energize.

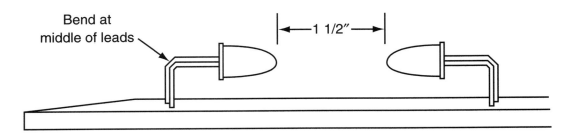

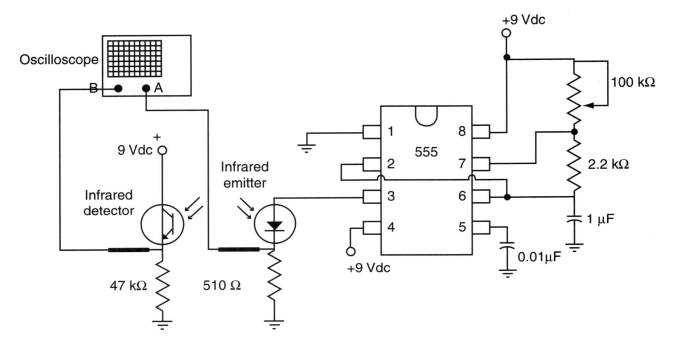

Step 3. Set the time/div. on the oscilloscope to at or about 2 ms and the volts/div. to 1 or 0.1 depending on if you are using a ×10 probe. Connect Channel A probe of the oscilloscope to the emitter circuit as indicated in the schematic.

Step 4. Energize the circuit with 9 Vdc.

Step 5. Adjust the 100 kΩ potentiometer at the 555 timer until a clear square wave appears on the oscilloscope generated at the emitter. There should be approximately four square waves across the screen of the oscilloscope. If there is not, try reversing the leads (polarity) to the emitter. Try also substituting a standard red LED. A standard red LED should shine bright or blink, depending on the setting of the 100 kΩ potentiometer. The red LED will verify that the timer circuit is working. If you still have a problem, check with your instructor.

Step 6. Connect Channel B probe of the oscilloscope to the infrared detector circuit as indicated in the schematic. A square wave should be observed on the scope; if not, try reversing the polarity to the detector.

Question 1. Compare the two square wave patterns displayed on the scope. Describe the similarities and the differences of the two square wave patterns.

_____

_____

Question 2. When the circuit is working properly and the devices have correct bias, what is the voltage drop across the following devices?

510 Ω resistor = _____ volt drop

Infrared emitter = _____ volt drop

Infrared detector = _____ volt drop

47 kΩ resistor = _____ volt drop

Question 3. When the circuit is not working properly due to reverse bias of the emitter and detector, what is the voltage drop across the following devices?

510 Ω resistor = _____ volt drop

Infrared emitter = _____ volt drop

Infrared detector = _____ volt drop

47 kΩ resistor = _____ volt drop

Step 7. Move the 3″ × 5″ card in and out of the beam between the emitter and detector several times while observing the oscilloscope.

Question 4. Describe the effect the card had on the infrared beam and the oscilloscope pattern.

_____

_____

Step 8. Take the small piece of aluminum foil and wrap it around the outside of the 2″ drinking straw. Connect the emitter and detector together using the 2″ drinking straw covered with foil.

Question 5. How did using the drinking straw to connect the light path from the emitter to the detector affect the system?

_____

_____

Step 9. Replace the infrared LED with the yellow LED, then the green LED, and finally the red LED.

Question 6. Describe the effect the yellow, green, and red LEDs had on the circuit, if any.

_____

_____

Step 10. Remove the 2″ drinking straw from the emitter and detector.

Step 11. Connect the 12-volt lamp to the 9-volt source using sufficient lengths of wire to bring the lamp near the detector when lit.

Question 7. Describe what happened to the detector signal on the oscilloscope when the incandescent lamp was brought near it.

_____

_____

Question 8. Do you think that a standard incandescent lamp produces any infrared light and why?

_____

_____

Step 12. Clear your work area. Properly store equipment and supplies.

25

Computers

Student Activity Sheet 25-1

Review

Name_____ **Score**_____

Date_____ **Class/Period/Instructor**_____

Complete the following sentences by filling in the missing word or words.

1. The Mark I was a(n) _____ calculator.

1. _____

2. The first electronic computer that used vacuum tubes was called _____.

2. _____

3. The second generation of computers used _____ rather than vacuum tubes and the third generation of computers uses _____ circuits.

3. _____

   _____

4. A microprocessor contains a complete _____ _____ on a single IC chip.

4. _____

5. The letters BASIC stand for _____ _____ _____ _____ _____.

5. _____

   _____

6. BASIC is a form of computer _____.

6. _____

7. The letters VLSI stand for _____ _____ _____ _____.

7. _____

   _____

8. The four basic parts of most computer systems are _____, _____, _____, and _____.

8. _____

   _____

   _____

   _____

9. An input or output device that is separate from, but connected to, the computer is called a(n) _____ device.

9. _____

10. Information on a data disk is stored as a series of _____ pulses that represent _____ and _____.

10. _____

    _____

    _____

11. The speed at which a modem transfers information is called its _____ rate.

11. _____

(Continued)

12. The term modem comes from the two words _____ and _____. 12. _____

_____

13. The letters CPU stand for _____ _____ _____. 13. _____

14. The cache is a(n) _____ _____ storage unit. 14. _____

15. The _____ unit is the brain of the CPU. 15. _____

16. The letters ALU stand for _____/_____ _____. 16. _____

17. Volatile memory is _____ when power to the computer is turned off. 17. _____

18. The letters ROM stand for _____ _____ _____ and the letters RAM stand 18. _____
for _____ _____ _____.

_____

19. The _____ memory cannot be erased but the _____ memory can be 19. _____
erased.

_____

20. The letters PROM stand for _____ _____ _____ _____. 20. _____

_____

21. A laser diode can be used to read data from a(n) _____. 21. _____

22. The _____ circuit synchronizes the operations of all the different parts of 22. _____
the computer.

23. The scanning device uses a(n) _____ to change light intensity into 23. _____
electrical pulses.

24. The letters GUI stand for _____ _____ _____. 24. _____

25. A LAN is a(n) _____ _____ _____ while a WAN is a(n) _____ _____ _____. 25. _____

_____

26. The computer that provides information and control of a LAN system is 26. _____
called the _____ _____.

27. A software program that is designed to destroy data and programs is 27. _____
called a(n) _____.

Temperature Display—Using the Keyboard as an Input Device

Name_____ **Score**_____

Date_____ **Class/Period/Instructor**_____

Introduction

In this laboratory activity, you will write a simple program in QBasic program which will turn the keyboard into an input system with the results displayed on the computer screen. The keyboard will be used as a temperature device which is being monitored by the computer. As the temperature is raised or lowered, it will be displayed on the screen. When the temperature is too high to be safe, an alarm will sound, alerting the operator of a problem.

Materials and Equipment

(1)—IBM PC or compatible

(1)—MS DOS Software System (standard with most computers) with QBasic.Exe

Procedure

Step 1. Boot up your computer.

Step 2. If you are in Windows, go to the Program Manager group and double-click on the MS DOS Prompt.

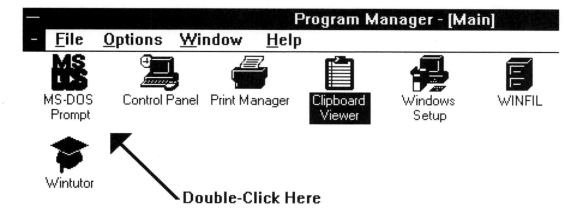

Step 3. The next display should look like

C:\WINDOWS>

Step 4. Type **QBASIC** and hit the return key.

Step 5. The display should appear like the one that follows.

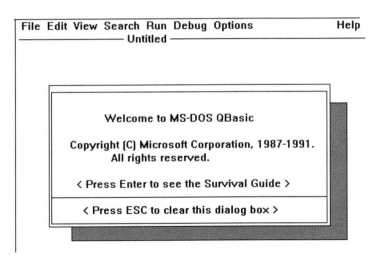

Step 6. Press the ESC key to clear the dialog box from the screen. Then, start typing the lines that appear below exactly as they appear. Spaces and punctuation should be exact.

common shared x, y, z

screen 9, 0

color 1, 7

x = 100

y = -40

line (300, 99)-(350, 250), 5, bf

line (299, 99)-(351, 251), 2, b

message

Step 7. Now use the mouse to drop down the menu under the Word File at the top left side of the screen.

Step 8. Choose "Save As" and save the program as "Temp" which is short for temperature. You have now saved the first module of the program. If you have been successful so far, the appearance of the program you have written thus far has changed some of the letters on the screen to capitals after you hit the return or enter key. The words with capitals are usually commands in QBasic.

Step 9. Select the "Edit" choice from the top left side of the screen by placing the mouse over it and clicking. A menu of selections should appear. Now click on "New Sub" and a dialog box will appear on the screen prompting you to type in the name of the new sub. Type in the name box "MESSAGE." Message will be the name of the subroutine that you will now type.

Step 10. A new screen should appear with the words "SUB Message" at the top left of the screen.

Step 11. Type the following now exactly as it appears below:

locate 2, 20: print "This is a data acquisition simulation.";

locate 3, 19: print " The keyboard will be the input device.";

locate 8, 46: print "200";

locate 10, 46: print "180";

locate 12, 46: print "160";

locate 14, 46: print "140";

Electronic Communication and Data Systems

```
locate 16, 46: print "120";

locate 18, 46: print "100";

while ansi$ <> "u" and ansi$ <> "U" and ansi$ <> "d" and ansi$ <> "D" and ansi$ <> "q"
and ansi$ <> "Q"

locate 22,1: input "Press u or d to raise and lower the temperature or q to quit", ansi$

wend

z = x - y

if z < 0 then z = 0

if z = 0 then x = 30

if z = 0 then y = 30

if z > 135 then z = 140

if z = 140 then x = 100

if z = 140 then y = -40

if z = 0 then locate 8,55: print "ALARM!!!!!!!!!!!";

if z = 0 then BEEP

if z <> 0 then locate 8,55: print "          ";

line (300, 100) - (350, 250), 5, bf

line (299, 99) - (351, 251), 2, b

line (300, 100 + z)-(350, 100 + (z + 10)), 0, bf

line (300, 100) - (350, 100 + z), 5, bf

line (250, 10) - (270, 260), 0, bf

line (250, z + 100) - (270, z + 100), 2

line (260, z + 110) - (270, z + 100), 2

select case ansi$

case "U", "u"

y = y + 10

x = x - 10

Message

case "D", "d"

x = x + 10

y = y - 10

Message

end

Message

end select
```

Step 12. Select file from the top left of the screen, and then click on "save."

Step 13. Press the F5 key, and watch the screen. A message about running the program should appear on the screen along with a blank temperature scale. Simply type the letter "u" and press "enter" or "return," and watch the display. An arrow and a line across the scale should appear and increment up each time you choose the letter "u." Pressing "d" and then the return or enter key will cause the temperature indicator to go down. You can press "q" to quit the program at anytime. When the temperature goes above 180 degrees, an alarm is indicated on the screen along with a "beep" sound.

In an industrial environment, a temperature transducer would be wired into one of the ports on a computer. The temperature transducer would convert the analog signal from a thermocouple to a digital signal that could be interpreted by the computer program. The temperature could be displayed on the screen. An alarm signal would go off if the temperature exceeded 180 degrees and could also shut down the electrical system to the equipment producing the heat. The system you programmed uses the keyboard as the transducer. Inputting the lowercase or uppercase letters u, d, or q will produce a response from the computer while ignoring all other letters.

Step 14. Now we will take a closer look at the program.

The first program is the main program called Temp.

DECLARE SUB Message ()	This statement appears automatically in the program after the subprogram has been named.
common shared x, y, z	Tells the program to share the variables x, y, and z.
screen 9, 0	Sets the resolution of the screen and the color.
color 1, 7	
x = 100	Sets the variables x and y to mathematical values to start with.
y = -40	
line (300, 99) - (350, 250), 5, bf	Draws the first outline of the thermometer
line (299, 99) - (351, 251), 2, b	
message	Tells the program to go to the subroutine called message.

Below is the subroutine called message:

locate 2, 20: print "This is a data acquisition simulation.";

locate 3, 19: print " The keyboard will be the input device.";

locate 8, 46: print "200";	The command word locate puts the cursor at the coordinates that follow (such as 8, 46 which means the row 8, column 46).
locate 10, 46: print "180";	
locate 12, 46: print "160";	The command word print will actually print whatever is in the quote marks.
locate 14, 46: print "140";	
locate 16, 46: print "120";	
locate 18, 46: print "100";	

Below the term while and wend is a control loop. The program will loop until a key is chosen, but only the keys u, d, or q.

while ansi$ <> "u" and ansi$ <> "U" and ansi$ <> "d" and ansi$ <> "D" and ansi$ <> "q" and ansi$ <> "Q"

locate 22, 1: input "Press u or d to raise and lower the temperature or q to quit", ansi$

wend

Below is a set of conditions for the variable called z. The variable z is a product of the variables x and y. The condition for the alarm is also set and the command word beep will produce a beep sound when z is equal to zero.

z = x - y

if z < 0 then z = 0

if z = 0 then x = 30

if z = 0 then y = 30

if z > 135 then z = 140

if z = 140 then x = 100

if z = 140 then y = -40

if z = 0 then locate 8,55: print "ALARM!!!!!!!!!!!";

if z = 0 then BEEP

if z <> 0 then locate 8,55: print "   ";

Below the line command word is used to draw lines and boxes which will represent the thermometer.

line (300, 100) - (350, 250), 5, bf

line (299, 99) - (351, 251), 2, b

line (300, 100 + z)-(350, 100 + (z + 10)), 0, bf

line (300, 100) - (350, 100 + z), 5, bf

line (250, 10) - (270, 260), 0, bf

line (250, z + 100) - (270, z + 100), 2

line (260, z + 110) - (270, z + 100), 2

Below is the special command called select case. It causes the program to loop between the words select case and the word message which is the name of the subroutine. It calls the subroutine over and over and looks for which key has been entered into the computer. The key selected will then cause the variables x and y to meet the mathematical condition stated. The mathematical sum totals for x, y, and z change each time a key is selected and will, in turn, cause the display of the thermometer to change. Look up at the command word line and you can see that the line draw is based upon the sum totals of x, y, and z.

select case ansi$

case "U", "u"

y = y + 10

x = x - 10

Message

case "D", "d"

x = x + 10

y = y - 10

Message

end

Message

end select

As you can see, each part of the program has a specific purpose. Some parts of the program interact and exchange information with each other. This program is meant to serve as an illustration of how a system can operate. There are over 150 commands in the QBasic language. QBasic is one of the latest versions of basic computer language. It may take many months, and in some cases years, to become proficient in the BASIC programming techniques. It is a real advantage for anyone interested in a career field in computers and electronic control systems to study some computer language, or better yet, several languages.

Step 15. Clear your area and properly shut down your computer.

Student Activity Sheet 25-3

Computer Bus Systems

Name_____ **Score**_____

Date_____ **Class/Period/Instructor**_____

Introduction

This laboratory activity is designed to represent the function of the typical computer bus system. The bus system is used to distribute binary codes through the data, control, and memory bus lines to components on the mother board and associated devices. A computer with only 16 bus lines can control thousands of components or place a high/low signal condition in thousands of memory locations. In this experiment, you will control four LEDs using only two conductors. Later, you will design a system with three conductors and determine the maximum number of devices that can be controlled by the three lines using digital techniques. This laboratory activity will provide you with a valuable insight to how a computer bus system operates.

Materials and Equipment

(1)—5-Vdc power supply

(4)—LEDs

(4)—510 Ω resistors, 1/4 W

(2)—SPST switches

(1)—breadboard

—digital devices required (to be determined by student)

Procedure

In the first part of this experiment, you will design a circuit to match the truth table that follows. There are four possible input states for the two single-pole switches—S_1 and S_2 open, S_1 closed and S_2 open, S_1 open and S_2 closed, both S_1 and S_2 closed. Each LED corresponds to a possible switch condition. Your job is to design a circuit that will meet each condition. You can use such devices as AND, OR, NOT, and NOR for example.

Binary Inputs		Switch Position		LED A	LED B	LED C	LED D
0	0	open	open	H	L	L	L
1	0	closed	open	L	H	L	L
0	1	open	closed	L	L	H	L
1	1	closed	closed	L	L	L	H

Step 1. Before gathering the materials, determine the type and number of digital devices you will need to accomplish the task. To do this, create a schematic of your circuit in the space that follows. Use this schematic to determine the devices you will need. Have your instructor approve your

schematic before assembly.

Step 2. *Gather required materials and assemble your circuit.* Do *not* energize until your instructor has checked your work.

Step 3. Demonstrate your circuit to your instructor for approval.

Step 4. Now design a similar circuit using three bus lines for control. Determine how many devices could be controlled. Draw the circuit below.

Question 1. How many devices can be controlled by three bus lines? _____

Question 2. Determine a formula for determining the number of devices controlled by the number of bus lines. Look for a pattern in the lab activity of two bus lines and three bus lines. Once the pattern is spotted, a simple math formula can be determined.

_____

Question 3. How many binary codes are represented by 4 bus lines? _____ How many codes are represented by 8 bus lines? _____ How many codes are represented by 16 bus lines? _____

Question 4. Make a truth table for a four-conductor bus on another sheet of paper.

Step 5. Clear your work area. Properly store equipment and supplies.

26

Career Opportunities in Electronics

Student Activity Sheet 26-1

Job Applications

Name _____ **Score** _____

Date _____ **Class/Period/Instructor** _____

    Below is a simulated typical job application. You will complete the application to the best of your ability. Be sure to neatly print all your information in the application. Remember, this document will be the first example your prospective employer will see of your ability to follow instructions and it will be a sample of your communication skills. This exercise will prove to be a valuable practice session for your future.

GW Electronic Enterprises Inc.

Employment Application Form 10113   Please Print.

Last Name	First Name	MI	Social Security #

Address:

Street, Apt #	City	Zip Code

DOB Day/Month/Year

	(Area Code) Phone Number

What position are you applying for?

_____

(Continued)

Work Experience: (Most Recent First)

Job Title	Company Name	Pay Rate	Address	Phone

List any special equipment you can operate:

_____

_____

_____

_____

_____

_____

Do you possess a valid drivers license? _____ Yes _____ No

When can you start, if hired today?_____

What pay rate do you desire? _____

List Education Experiences: Include any Degrees, Diplomas or Certificates Received:

School	Location	Degree, Certificate or Diploma

References: (Include Most Recent Employer)

Name	Address	Phone

Briefly state why you desire employment with our firm.

_____

_____

_____

_____

_____

_____

_____

_____

_____

_____

_____

_____

_____

_____

_____

_____

_____

_____

_____

_____

_____

_____

_____